SpringerBriefs in Law

SpringerBriefs in Law and Regulation & Innovation

Kai P. Purnhagen
Chair of Food Law, University of Bayreuth
Kulmbach, Germany

SpringerBriefs present concise summaries of cutting-edge research and practical applications across a wide spectrum of fields. Featuring compact volumes of 50 to 125 pages, the series covers a range of content from professional to academic. Typical topics might include:

- A timely report of state-of-the art analytical techniques
- A bridge between new research results, as published in journal articles, and a contextual literature review
- A snapshot of a hot or emerging topic
- A presentation of core concepts that students must understand in order to make independent contributions

SpringerBriefs in Law showcase emerging theory, empirical research, and practical application in Law from a global author community. SpringerBriefs are characterized by fast, global electronic dissemination, standard publishing contracts, standardized manuscript preparation and formatting guidelines, and expedited production schedules.

SpringerBriefs in Law and Regulation & Innovation present summaries of cutting-edge research and practical applications across the wide spectrum of regulatory law. Regulation has established itself as an important component of the legal system. It aims to achieve the common good in the interaction between administration and private individuals and occurs alongside the other central (administrative) legal functions of limiting state power to protect freedom, averting dangers, regulating the direct provision of state services and the general provision of legal rules for social development. SpringerBriefs in Law and Regulation & Innovation are dedicated to the understanding, systematic recording and dogmatic processing of such regulatory regimes. It is open for basic questions, central elements and new developments in the regulatory law. The focus is on legal regimes to secure results that promote the common good from and through economic competition, such as those found in network economies and increasingly in the food as well as in the health sector etc. Featuring compact volumes of 50 to 125 pages, the series covers a range of content from mainly professional but also academic.

Alexandra Molitorisová • Alessandro Monaco
Gerald Lackner • Seth Roberts
Kai P. Purnhagen

Regulation of Mushroom- and Mycelium-Based Products in the EU

Springer

Alexandra Molitorisová
Chair of Food Law
University of Bayreuth
Kulmbach, Germany

Alessandro Monaco
Chair of Food Law
University of Bayreuth
Kulmbach, Germany

Gerald Lackner
Biochemistry of Microorganisms
University of Bayreuth
Kulmbach, Germany

Seth Roberts
Good Food Institute Europe
Brussels, Belgium

Kai P. Purnhagen
Chair of Food Law
University of Bayreuth
Kulmbach, Germany

ISSN 2192-855X ISSN 2192-8568 (electronic)
SpringerBriefs in Law
ISSN 3004-9210 ISSN 3004-9229 (electronic)
SpringerBriefs in Law and Regulation & Innovation
ISBN 978-3-031-93779-8 ISBN 978-3-031-93780-4 (eBook)
https://doi.org/10.1007/978-3-031-93780-4

This Springer imprint is published by the registered company Springer Nature Switzerland AG
The registered company address is: Gewerbestrasse 11, 6330 Cham, Switzerland

Acknowledgements

This book is a significantly updated and expanded version of the report "Analysis of the regulatory environment applicable to products obtained from mushrooms and fungal mycelium", published in 2021. The report was the outcome of a research project of the Chair of Food Law of the Faculty of Life Sciences: Food, Nutrition and Health (Campus Kulmbach) and the Forschungsstelle für Deutsches und Europäisches Lebensmittelrecht of the University of Bayreuth, with the financial support of the Adalbert-Raps-Stiftung, the Deutsche Forschungsgemeinschaft, project number 465588286, and the Oberfrankenstiftung.

We thank Frank Kühne, Sebastian Sommerer, Elisabeth Lagenberger and the Adalbert-Raps-Stiftung for their trust and the opportunity to conduct this research, which continues to offer new insights and inspiration.

We thank those who believed in the project and helped us in our research by providing useful contacts: Alex Holst (The Good Food Institute Europe), Frank Anders (NX-Food), and Albrecht Wolfmeyer (ProVeg).

FLMR Forschungsstelle für Deutsches und Europäisches Lebensmittelrecht der Universität Bayreuth

UNIVERSITÄT BAYREUTH

Chair of Food Law
Lehrstuhl für Lebensmittelrecht

Funded by

DFG Deutsche Forschungsgemeinschaft
German Research Foundation

Competing Interests The authors declare the following as potential competing interests: Seth Roberts is Policy Manager of The Good Food Institute Europe.

Key Terms and Summary

Key Terms

Agricultural products are products of the soil, of stock farming and of fisheries and products of first-stage processing directly related to these products falling under one of the categories of Annex I of the Treaty on the Functioning of the European Union.

Authorization procedures are a tool used for pre-market approval of substances and products which are deemed to present potential risks for consumer health and interests and for the environment. They are based on risk analysis and, in the European Union, consist of two important steps: an independent risk assessment by scientific authorities and a political decision of risk management taken by political authorities.

Claim is any message or representation, which is not mandatory under Community or national legislation, including pictorial, graphic or symbolic representation, in any form, which states, suggests or implies that a food has particular characteristics.

Clean labels are labels which reflect the consumers' rejection of highly processed products and, in their broadest sense, represent the changes in perception which affected consumers, particularly in developed countries, and the increased interest for healthy, natural and sustainable products, both for the people and for the planet.

Environmental claims are any message or representation which is not mandatory under Union or national law, in any form, including text, pictorial, graphic or symbolic representation, such as labels, brand names, company names or product names, in the context of a commercial communication, and which states or implies that a product, product category, brand or trader has a positive or zero impact on the environment or is less damaging to the environment than other products, product categories, brands or traders, or has improved its impact over time.

Feed means any substance or product, including additives, whether processed, partially processed or unprocessed, intended to be used for oral feeding to animals.

Fermentation refers to the conversion of substrates (like sugars) into desirable products (such as alcohol, acids or gases) by microorganisms during the production of foods or beverages.

Food (or "foodstuff") means any substance or product, whether processed, partially processed or unprocessed, intended to be, or reasonably expected to be ingested by humans.

Food additive means any substance not normally consumed as a food in itself and not normally used as a characteristic ingredient of food, whether or not it has nutritive value, the intentional addition of which to food for a technological purpose in the manufacture, processing, preparation, treatment, packaging, transport or storage of such food results, or may be reasonably expected to result, in it or its by-products becoming directly or indirectly a component of such food.

Food enzymes are products obtained from plants, animals or micro-organisms or products thereof, including a product obtained by a fermentation process using micro-organisms, that contains one or more enzymes capable of catalysing a specific biochemical reaction and are added to food for a technological purpose.

Food information means information concerning a food and made available to the final consumer by means of a label, other accompanying material, or any other means including modern technology tools or verbal communication.

Food supplements are defined as foodstuffs the purpose of which is to supplement the normal diet and which are concentrated sources of nutrients or other substances with a nutritional or physiological effect, alone or in combination, marketed in dose form, namely forms such as capsules, pastilles, tablets, pills and other similar forms, sachets of powder, ampoules of liquids, drop dispensing bottles, and other similar forms of liquids and powders designed to be taken in measured small unit quantities.

Filamentous fungi are fungal species that form hyphae, thread-like, branching structures, which permeate the growth substrate and form a reticulate network called mycelium. They comprise the moulds and mushrooms but exclude unicellular fungi (yeasts).

Flavourings are products: (i) not intended to be consumed as such, which are added to food in order to impart or modify odour and/or taste; (ii) made or consisting of the following categories: flavouring substances, flavouring preparations, thermal process flavourings, smoke flavourings, flavour precursors or other flavourings or mixtures thereof.

Fruiting body is the reproductive structure of the fungal organism, normally growing above the ground, which is, in edible species, consumed by humans.

Fungi are organisms such as yeasts, moulds, and mushrooms belonging to the biological kingdom Fungi, a group distinct from animals, plants and bacteria.

Green claims are assertions made by producers, traders, manufacturers and retailers about the environmentally beneficial aspects of their products.

Health claims are claims that states, suggests or implies that a relationship exists between a food category, a food or one of its constituents and health.

Label means any tag, brand, mark, pictorial or other descriptive matter, written, printed, stencilled, marked, embossed or impressed on, or attached to the packaging or container of food.

Medicinal product means any substance or combination of substances presented as having properties for treating or preventing disease in human beings or any substance or combination of substances which may be used in or administered to human beings either with a view to restoring, correcting or modifying physiological functions by exerting a pharmacological, immunological or metabolic action, or to making a medical diagnosis.

Mushrooms are fruiting bodies of fungi, typically consisting of a stalk and cap. Most mushroom-forming fungi, also commonly referred to as "mushrooms" themselves, belong to the phylum *Basidiomycota,* which includes the common edible mushrooms.

Mycelium is an interconnected web of hyphae (filamentous, thread-like structures) that constitutes the vegetative body of the fungus, accessing resources and growing through its environment.

Novel food in the European Union means any food that was not used for human consumption to a significant degree within the Union before 15 May 1997, irrespective of the dates of accession of Member States to the Union and which can be categorized in one of the novel food categories.

Nutrition claim means any claim which states, suggests or implies that a food has particular beneficial nutritional properties due to: (a) the energy (calorific value) it (i) provides; (ii) provides at a reduced or increased rate; or (iii) does not provide; and/or (b) the nutrients or other substances it (i) contains; (ii) contains in reduced or increased proportions; or (iii) does not contain.

Organic production means the use, including during the conversion period, of production methods that comply with this Regulation at all stages of production, preparation and distribution. Any farm that wishes to produce organically must undergo a process known as "conversion", during which production methods need to be used.

Organic product is a product resulting from organic production, other than a product produced during the conversion period. The products of hunting or fishing of wild animals are not considered as organic products.

Petfood is the feed of non-food producing animals.

Substrate is the material on which an organism grows and from which it derives its nutrients.

Yeasts are fungi that mainly occur as single cells and undergo cell division by budding or fission as a primary means of asexual reproduction. Their sexual stages do not develop in fruiting bodies.

Summary

This book provides a detailed account of food law provisions applicable to mushroom and mycelium products (MMP) produced or placed on the market in the European Union (EU). The objective of the book is to map the regulatory environment governing mushrooms and mycelium products, focusing on the pre-market authorisation and labelling requirements.

The first section of the book concerns the classification of MMP. Whether MMPs are categorized as foods, food supplements, food additives, flavourings or enzymes depends on their intended use. MMP may also fall under the classification of medicinal products, cosmetics or petfood and feed, which would exclude them from food law regulations. Food business operators handling borderline products (for example, between foods and medicinal products) should carefully consider their claims and related pieces of evidence. MMP may also be subject to the provisions of the Common Agricultural Policy.

MMP used as foods can potentially be classified as novel foods. Mushroom species not commonly consumed in the EU, and products obtained from mycelium, even of species whose fruiting bodies have a long history of consumption, can be considered novel foods and are likely to be subject to Regulation (EU) 2015/2283 on novel foods ("Novel Food Regulation"). The Novel Food Regulation imposes stringent regulatory requirements on applicants seeking novel food authorizations. These requirements include demonstrating safety through comprehensive and robust scientific evidence, such as multiple toxicity studies using animal models or human data. In the market segment for MMPs, the authorization of biomass fermentation involving various fungal mycelia is particularly relevant.

The pioneering phase of the MMP sector is largely driven by start-ups, which often operate with limited financial resources. The high regulatory demands of the Novel Food Regulation can significantly impede innovation in this sector, forcing food business operators to prioritize regulatory compliance over other business or societal considerations in their strategies.

Rules on the labelling of MMP concern both mandatory and voluntary food information provided to consumers. The first issue to consider is the correct label designation for mushrooms and mycelium products. The preference for generic names over scientific names should be guided by how consumers commonly refer to these products. The use of meat-like terms to describe MMP products should be assessed on a case-by-case basis. For voluntary labelling, claims such as "plant-based" must be substantiated with appropriate evidence. Nutrition and health claims can be included provided they do not mislead consumers and comply with applicable legal requirements.

Food business operators involved in the production or marketing of organic products may seek certification from accredited bodies. Alternative protein products derived from MMP are often categorized as "vegan" or "vegetarian" and frequently feature sustainability claims. In the absence of specific EU or national regulations, private certification schemes can be employed to indicate whether products are vegan or vegetarian and to support the use of sustainability claims.

The regulatory landscape applicable to MMP within the EU is robust and evidence-based, but it does present some opportunities for efficiency and effectiveness improvement that could be addressed by policymakers. First, using agricultural by-products as substrates for fermentation could increase the sustainability of the industry but it is limited by food safety provisions and EU waste law. Provisions could be adopted to facilitate the use of specific products as substrates. Second, to increase the efficacy and efficiency of the risk assessment during the authorization procedure, new assessment methodologies, including the software-based prediction of biosynthetic gene clusters or the inclusion of fungi on the Qualified Presumption of Safety (QPS) list, could be used. Finally, labelling rules on the use of meat like denominations and wordings for alternative protein products could be clarified to facilitate the marketing of MMPs.

Contents

Abbreviations

CAP	Common Agricultural Policy
CJEU	Court of Justice of the European Union
EU	European Union
FAR	Food Additive Regulation
FIC	Food Information to Consumers Regulation (Regulation (EU) No 1169/2011)
GFL	General Food Law (Regulation (EC) No 178/2002)
MMP	Mushroom and Mycelium's product
NFR	Novel Food Regulation (Regulation (EU) 2015/2283)
TFEU	Treaty on the Functioning of the European Union

Chapter 1
Introduction, Background and Methodology of the Project

Abstract This chapter introduces the scope and objectives of the book. It begins with an overview of mushrooms and mycelium-based products, emphasizing their unique characteristics, morphological distinctions, and diverse applications within the food industry. The chapter establishes a foundation for examining their potential, as well as the challenges they encounter within the EU regulatory framework. This sets the stage for detailed discussions on regulations and market implications in the chapters that follow.

1.1 Introduction

In 2019, the European Union (EU) published the European Green Deal, a package of policies that is designed to make the Union climate neutral by 2050.[1] The production of animal proteins is one of the most environmentally impacting industries because it requires significant resources.[2] The reduction of animal proteins consumption is explicitly recognized as a critical step to achieve the objectives of the Farm-to-Fork (F2F) strategy, which the European Commission communicated in May 2020 as a strategic blueprint for food systems' transformation in the EU.[3]

The Commission has also committed to reviewing the EU protein policy to enhance food security while minimizing environmental and climate impacts both within the EU and globally.[4] For this reason, food and feed innovators have been focussing on the development and production of alternative proteins.[5] Multiple food

---

[1] European Commission (2019).

[2] Smetana et al. (2023).

[3] European Commission (2020).

[4] See the answer given by Mr Wojciechowski on behalf of the European Commission to a parliamentary question on 2 February 2023 P-000044/2023(ASW). https://www.europarl.europa.eu/doceo/document/P-9-2023-000044-ASW_EN.html.

[5] Rzymski et al. (2021).

A. Molitorisová et al., *Regulation of Mushroom- and Mycelium-Based Products in the EU*, SpringerBriefs in Law, https://doi.org/10.1007/978-3-031-93780-4_1

technologies, such as cultivated meat and precision fermentation,[6] are considered critical in the effort to transform protein production and consumption in the EU.

Products derived from mushrooms and fungal mycelium represent a valid alternative to proteins of animal origin. Humans have traditionally been consuming fruiting bodies of mushrooms. More recently, food innovators have focused on the unexplored opportunities offered by fungal mycelium.[7] The mycelium is the filamentous web which constitutes the vegetative body of the mushrooms and fungi. If cultivated under appropriate conditions, mycelium can serve multiple purposes.[8] Food innovations are driven by diverse objectives. Some aim to enhance food diversity or provide consumers with greater sensory enjoyment and pleasure, while others are designed to address critical societal challenges, such as reducing the climate impact of food production. Ideally, food innovations are scalable, allowing them to make a meaningful contribution to markets, consumers, and society at large.

Due to the inherent nature of food, many innovations focus on utilizing compounds found in plants, algae, or fungi. These innovations often involve repurposing familiar compounds for new applications, uncovering their potential through advancements in knowledge, techniques, processes, and substances.

Developers and producers of mushroom and mycelium products benefit from a remarkable biological diversity in the fungi kingdom. Many strains of fungi have been part of the human diet. However, this statement has a great geographical and temporal variability. In some parts of the world, a greater variety of fungi strains are eaten daily compared to others. In the past, eating wild mushrooms was more common than today. In different regions of the world, different biotechnological uses of fungi may be found, such as cultivation, fermentation, and others. Climate change impacts the availability of fungi in the natural environment in certain regions. All this impacts how fungi are perceived, used, and regulated in a particular market.

Fungi are decomposers of biomass in nature—of anything that is hydrocarbon-based. This means that fungi can be used for different decomposing purposes but also that for many people, fungi are associated with decay. Fungi are neither bacteria, nor plants. They are not animals either. They constitute their own distinct biological kingdom. Some say they represent the digestive tracks of nature. There are over 1.5 million fungi species, 6 times more than plants. Out of this great number of fungi species, many are poisonous and represent a threat to the life or health of human beings. Only a handful forms edible fruiting bodies, commonly known as mushrooms, and have been consumed safely, with the safe consumption to be diligently documented. From the long-standing interactions between fungi and humans, current food innovations emerge.

Viable spores or mycelium of a fungal strain are injected into a substrate full of nutrients and, under appropriate conditions, they modify the substrate's composition and/or turn it into fungal protein biomass. Fungal protein biomass can then be

[6] Ronchetti et al. (2024).

[7] Bayer (2019).

[8] Bayer (2019).

simply sliced and cooked as such, or mycelium-based proteins can be further processed into powders and extracts. Foods obtained through mycelium fermentation represent an alternative to animal proteins,[9] and upcycling of food waste often relies on mycelium's metabolic activities. Outside of the food sector, mycelium is also used as a substitute of plastic, leather and wood.[10]

As of 2024, the global market for alternative proteins is valued at €13.5 billion and is projected to grow 8.2% yearly until 2030,[11] with the largest segment still dominated by established meat alternatives, particularly soy-based products. The fungi market, valued at nearly €38 billion in 2019, is expected to expand at an average annual rate of 8.3% through 2027.[12] This analysis focuses on how these market opportunities can be harnessed within the food sector, considering the current regulatory framework governing products derived from mycelium in the EU.

1.2 Purpose of the Research

Applicable law and regulations have an influence on the successful adoption of food innovations in society. Regulation can either be viewed as an incentive for innovation,[13] or as an obstacle, particularly in the food sector.[14] If it shall work as an incentive, predictability which comes with legal certainty is essential. However, significant uncertainty regarding the regulatory framework in the EU for mushroom-based and mycelium-based products (MMPs) exists. This lack of legal certainty regarding which provisions apply to specific products constitutes a major problem for food business operators and stakeholders who wish to access the market and engage in the sector.

Throughout this book, we aim at mapping the EU regulatory environment applicable to food products obtained from mushrooms and mycelia, to offer stakeholders (e.g. companies, start-ups, investors and policy makers) the instruments to navigate the EU regulatory landscape. Additionally, the research highlights gaps in the legislation and its interpretation, which may serve as a starting point for future academic research and policies.

---

[9] Kumar et al. (2021).

[10] For a review of mycelium application in the food sector and beyond see Business Insider (2021a, b).

[11] EVE Grand View Research (2024).

[12] MarketWatch (2021).

[13] Porter (1991).

[14] Lähteenmäki-Uutela et al. (2021); Purnhagen Wesseler (2021).

1.3 Methodology

The chapters of the book are based on a doctrinal legal analysis, expert interviews and a literature review. Doctrinal legal analysis describes the collection and ordering of the available legal material, including legislation, jurisprudence, and legal literature.[15] The doctrinal legal research consisted of an in-depth analysis of relevant EU primary law, including the case law of the Court of Justice of the European Union, EU secondary law (the relevant regulations and directives), as well as the most relevant literature and soft law at the national, European, or international level. Each of these sources was employed to provide a clear understanding of legal requirements applicable to MMPs in the area of EU food law. Throughout the book, legal provisions are cited in footnotes but paraphrased in the main text to (1) offer legal advisors or companies' regulatory experts a collection of applicable legislation and (2) make legal provisions accessible to "non-lawyers". The doctrinal analysis was combined with interdisciplinary methods to assess the application of the law in practice and potential requirements for reform.[16] In this respect, we first merged doctrinal legal research with empirical insights garnered through expert interviews.[17] We conducted expert interviews with stakeholders active in the sector to collect first-hand accounts of regulatory hurdles applicable to food products obtained from mushrooms and mycelia. We also merged doctrinal legal analysis with insights gained from a literature review concerning toxicology studies.

1.4 Chapters of the Book

The Chapter "Classification of Mycelium and Mushroom Products" examines the legal requirements for placing food products derived from mushrooms and mycelium on the market in the European Union. It compares the definition of food with those of food additives, flavourings, food supplements, and food enzymes, highlighting the distinctions when applied to MMPs. Additionally, it explores the conditions under which mycelium products can be categorized as "agricultural products" under EU law.

The Chapter "Novel Food Status of Mycelium and Mushroom Products" provides a comprehensive analysis of the EU legal framework governing novel foods and pre-market authorization of MMPs. By considering several examples of consultations concerning the novel food status of MMPs, the Chapter considers the requirements to demonstrate a history of safe use. It then focuses on the technical aspects relevant to the preparation of a novel food application, considering, in particular, the most recent guidance documents published by the EFSA. The Chapter

[15] Hutchinson Duncan (2012).

[16] Hutchinson (2015).

[17] See for an application Warda et al. (2024); Purnhagen et al. (2021).

also critically examines the Qualified Presumption of Safety (QPS) list and other mechanisms aimed at streamlining the risk assessment process for MMPs. Additionally, it highlights several challenges and gaps in toxicological data related to filamentous fungi, offering insights into the complexities of ensuring safety for these products.

The Chapter "Labelling of Mushrooms and Mycelium Products" provides a comprehensive guide on the necessary information that should be conveyed to consumers regarding mushroom/mycelium-based products. It starts by delving into the specific naming conventions for mycelium products. It then shifts focus to the regulations applicable to claims made on the products, particularly the use of the "plant-based" label and of health and nutrition claims associated with MMPs. The Chapter also offers an overview of additional voluntary information that food businesses may choose to provide. This includes an exploration of the concept of clean labelling, an explanation of the criteria for vegan and vegetarian designations, a breakdown of the requirements for organic certification, and a discussion on the recent regulatory development over the use of environmental claims on food product labelling.

Instruments and Legislation

European Commission, Communication From the Commission to the European Parliament, the Council, the European Economic and Social Committee and the Committee of the Regions: A Farm to Fork Strategy for a fair, healthy and environmentally-friendly food system 2020 (COM(2020) 381 final)

European Commission, Communication on the European Green Deal 2019 (COM/2019/640)

References

Bibliography

Bayer E (2019) The Mycelium Revolution Is upon Us. Scientific American Blog Network, <https://blogs.scientificamerican.com/observations/the-mycelium-revolution-is-upon-us/>

Business Insider (2021a) How mushrooms are turned into bacon and styrofoam | World Wide Waste. <https://www.youtube.com/watch?v=uznXI8wrdag>

Business Insider (2021b) Mushroom coffins turn bodies into compost and make death less toxic | World Wide Waste, <https://www.youtube.com/watch?v=AurhO4Lf1Is>

EVE Grand View Research (2024) Alternative Protein Market Size, Share & Trends Report Alternative Protein Market Size, Share & Trends Analysis Report, By Product (Plant Protein, Microbe-Based Protein, Insect Protein), By Application (Food & Beverages, Infant Formulations, Clinical Nutrients, Animal Feed), By Region, And Segment Forecasts, 2024 – 2030, GVR-4-68040-197-4

Hutchinson T (2015) The doctrinal method: incorporating interdisciplinary methods in reforming the law. Erasmus Law Rev 3:130. https://doi.org/10.5553/ELR.000055

Hutchinson T, Duncan N (2012) Defining and describing what we do: doctrinal legal research. Deakin Law Rev 17:83. https://doi.org/10.21153/dlr2012vol17no1art70

Kumar K, Mehra R, Guiné RPF, Lima MJ, Kumar N, Kaushik R, Ahmed N, Yadav AN, Kumar H (2021) Edible mushrooms: a comprehensive review on bioactive compounds with health benefits and processing aspects. Foods 10:2996. https://doi.org/10.3390/foods10122996

Lähteenmäki-Uutela A and others (2021) Alternative proteins and EU food law. Food Control 130:108336. https://doi.org/10.1016/j.foodcont.2021.108336

MarketWatch (2021) Mushroom Market - Global Industry Analysis By Development, Size, Share and Demand Forecast, <https://www.marketwatch.com/press-release/mushroom-market%2D%2D-global-industry-analysis-by-development-size-share-and-demand-forecast-2021-07-02>

Porter ME (1991) America's green strategy. Sci Am 264:168

Purnhagen K, Wesseler J (2021) EU regulation of new plant breeding technologies and their possible economic implications for the EU and beyond. Appl Econ Persp Policy 43:1621

Purnhagen K, Dries L and Preusse V (2021) The Brussels effect in action: do the EU's standards travel abroad? A study on control systems for geographical indications in the EU and Thailand. Houston J Int Law 43:455

Ronchetti F, Springer L, Purnhagen K (2024) The regulatory landscape in the EU for dairy products derived from precision fermentation. Springer Briefs in Law, https://doi.org/10.1007/978-3-031-49692-9

Rzymski P and others (2021) COVID-19 pandemic is a call to search for alternative protein sources as food and feed: a review of possibilities. Nutrients 13:150

Smetana S and others (2023) Meat substitutes: resource demands and environmental footprints. Resour Conserv Recycl 190:106831

Warda R, Purnhagen K, Molitorisová M (2024) Has mutual recognition in the EU failed?—A legal-empirical analysis on the example of food supplements containing botanicals and other bioactive substances. J Consumer Policy 47:425

Chapter 2
Classification of Mycelium and Mushroom Products

Abstract The Chapter examines the legal requirements for placing food products derived from mushrooms and mycelium on the market in the European Union. It compares the definition of food with the ones of food additives, flavourings, food supplements, and food enzymes, highlighting the differences when applied to MMPs. Additionally, it explores the conditions under which mycelium products can be categorized as "agricultural products" under EU law.

2.1 The Definition of Food in EU Law

Within the European Union (EU), the production and commercialization of goods classified as "food" are strictly regulated by a comprehensive set of legislative acts collectively known as Food Law.[1] Products are classified as "foods" based on their intended use, which is determined by observable objective criteria. Regulation (EC) No. 178/2002 (the General Food law or GFL), defines "food" as meaning "any substance or product, whether processed, partially processed or unprocessed, intended to be, or reasonably expected to be ingested by humans."

When products are classified as foods, they can be lawfully placed on the market without further notice, and their trade cannot be restricted within the Union, providing that they:

(1) Do not fall within a separate regulatory framework, e.g. the regulatory framework applicable to medicinal products.
(2) Comply with specific food law provisions requiring a pre-market authorisation, e.g. Regulation (EU) No 2283/2015 on novel foods or Regulation (EC) No 1829/2003 on genetically modified food.

[1] Regulation (EC) No 178/2002, Article 3.

A. Molitorisová et al., *Regulation of Mushroom- and Mycelium-Based Products in the EU*, SpringerBriefs in Law, https://doi.org/10.1007/978-3-031-93780-4_2

(3) Comply with general food law requirements, e.g. food is not injurious to health or unfit for human consumption,[2] and is labelled and traced.[3]
(4) Comply with any other relevant European or national legal provision that might be applicable in specific situations.[4]

Whenever these conditions are met, Mushrooms and Mycelium Products (MMP) can be placed on the market as foods in the EU.

2.2 What Are MMP?

Mushroom and mycelium products are made from filamentous fungi. Therefore, before discussing the safety of filamentous fungi, it is important to revisit the definitions and scope of the label "filamentous fungi". Fungi are organisms belonging to the taxonomic kingdom Fungi. Important members of the Fungi are commonly known as yeasts, moulds and mushrooms, none of which represent taxonomic units according to scientific standards. Filamentous fungi, which comprise moulds and mushrooms, are fungal species that form filaments (hyphae). Hyphae (singular: hypha) are thread-like, branching structures, which permeate the growth substrate or extend into the air. The mesh-like entirety of hyphae is referred to as the mycelium. Highly interconnected (anastomosed) and dense mycelia can also aggregate into tissue-like structures. Although they are not tissues (closed layers of cells) in a strict sense, they are able to build up morphologically complex structures like fruiting bodies. From a taxonomic point of view, the "Fungi" represent a kingdom, a high-level classification that places them as an independent group within the tree of life, distinct from bacteria, plants and animals. Taxonomists divide fungi into divisions, most importantly the *Mucormycota* (informally called zygomycetes), *Ascomycota* (ascomycetes) and *Basidiomycota* (basidiomycetes).[5] These phylogenetic categories are central to fungal biology and their members exhibit not only distinct genetic and morphological traits, but also fundamentally different reproductive cycles. Here, only the most important concepts are summarized.

Notably, filamentous fungi are present across all major groups of fungi. Important members of the zygomycetes, for instance, are *Rhizopus oligosporus* or *R. oryzae*, mould species used for the fermentation of soybeans to tempeh.[6] During sexual reproduction, they form bridge-like structures between hyphae of compatible mating types. In the centre of these zygosporangia, oval zygospores are formed, which often exhibit thick and heavily ornamented walls. A hallmark of ascomycetes, however, is the formation of sexual spores within sheath-like structures, called asci,

[2] Regulation (EC) No 178/2002, Article 14.
[3] Regulation (EC) No 178/2002, Articles 16, 18.
[4] Regulation (EC) No 178/2002, Article 14(9).
[5] Spatafora et al. (2017).
[6] Londoño-Hernández et al. (2017); Tian et al. (2022).

which are found within fruiting bodies. Ascomycetes comprise typical filamentous moulds like *Aspergillus* and *Penicillium* species, which play a role not only in food production (e.g. blue cheese ripening) but frequently occur as dangerous, toxin-producing spoilage organisms. These mould species within the ascomycetes belong to the subdivision *Pezizomycotina*. Confusingly, many mould species do not undergo any sexual life cycle involving ascus formation under laboratory conditions. Because of the lack of a sexual form (teleomorph) and only existing in an asexual form (anamorph) they are traditionally called *fungi imperfecti*. Members of the *Pezizomycotina* subdivision are especially prolific producers of bioactive compounds.

Basidiomycetes are another important group of filamentous fungi. They develop sexual spores on tiny horns that extend from pedestals called basidia. These tiny structures are not visible by eye and occur, for instance, in the spore forming parts of the fruiting bodies underneath the mushroom cap. The mushroom itself is a fruiting body, typically consisting of a stalk and cap. However, the mushroom-forming organism itself is also often referred to as a mushroom. Prominent examples are the button mushroom (*Agaricus bisporus*), the oyster mushroom (*Pleurotus ostreatus*) or the shiitake (*Lentinula edodes*), all basidiomycetes. Numerous mushrooms are also famous for toxin production. For example, the death cap (*Amanita phalloides*) is notorious for producing amanitins, cyclic peptides that cause fatal liver failure when ingested.[7] While mushroom-forming basidiomycetes represent an important class of filamentous fungi, basidiomycetes might also appear in the shape of molds or even yeasts.

Contrary to filamentous fungi, yeasts are described as fungi that mainly occur as single cells and undergo cell division by budding or fission as a primary means of asexual reproduction. Their sexual stages do not develop in fruiting bodies.[8] Most yeasts are members of the ascomycetes but belong to the subdivisions *Saccharomycotina* (e.g. the baker's yeast *S. cerevisiae*) or *Taphrinomycotina* (e.g. the fission yeast *Schizosaccharomyces pombe*), never to the *Pezizomycotina*, which are typically filamentous. This definition of yeasts comprises certain dimorphic fungi, which are able to switch between filamentous and unicellular growth, such as *Candida albicans*. Dimorphic fungi belonging to the *Pezizomycotina*, are not regarded as yeasts. As previously noted, a few yeast species also belong to the basidiomycetes. The definition of yeast is particularly significant, as yeasts are generally considered suitable for EFSA's QPS process.

In summary, filamentous fungi represent a phylogenetically heterogeneous group, encompassing molds from zygomycetes and ascomycetes (especially the *Pezizomycotina*) as well as the basidiomycetes.

---

[7] Luo et al. (2022).

[8] Kurtzman et al. (2011).

2.3 Difference Between Mycelium and Fruiting Body

Fruiting bodies (also: fruit bodies or carpophores) are spore-bearing structures of higher fungi (ascomycetes and basidiomycetes), which are formed during the sexual reproductive lifecycle. Typical fruiting bodies, especially of basidiomycetes, are "mushrooms", characterized by a cap and stalk. However, fruiting bodies can also adopt different forms, resembling for instance shelves (conks), corals or cups. A few members of the ascomycetes also produce edible fruiting bodies (ascocarps), such as the highly valued morels and truffles. In general, fruiting bodies are composed of tightly interwoven and interconnected (anastomosed) hyphae, forming tissue-like structures that develop specialized formations for spore production. Ascomycetes produce "asci" within their ascocarps, tube-like structures that typically contain eight ascospores. Basidiomycetes form "basidia", which bear four basidiospores each, sitting on tiny stalks the protrude from the basidium. Basidia are found, for instance, within lamellae (gills) or pores underneath the mushroom cap. The powder found under the mushroom consists of liberated basidiospores.

The formation of fruiting bodies from vegetative mycelia is a complex developmental program, which requires certain prerequisites. Filamentous mushrooms typically initiate their growth as vegetative mycelia, which permeate the substrate, such as soil, decaying wood, or artificial growth media. During the sexual life cycle, the formation of a fruiting body is preceded by the mating of mycelia from two compatible mating types. During the fusion of hyphae (plasmogamy), a dikaryotic mycelium is formed, containing two genetically distinct nuclei per cell. When certain internal (e.g., nutrient accumulation) and external conditions (e.g., nutrient depletion, temperature, humidity, light) are met, fruiting bodies may be formed, a process known as fructification. Fructification begins with the formation of small knots or hyphal aggregates, called primordia. These develop further into fruiting bodies, where spore formation and release ultimately take place.

Mold-like fungi such as *Aspergillus* or *Penicillium* species, which belong to the ascomycetes, do not form mushrooms. However, they do form different developmental structures, whose development is often linked to the regulation of bioactive metabolite biosynthesis.[9] Ascomycetes often lack sexual reproductive stages (teleomorphs) and primarily reproduce via conidia (asexual spores) formed on specialized spore-bearing structures called conidiophores. Conidia are the well-known spores released by mould fungi causing food spoilage. While some ascomycete mold species can reproduce sexually, their fruiting bodies, known as cleistothecia, are not relevant to food production. More significant are the sclerotia—resilient structures produced by certain fungal genera—which are often linked to the production of bioactive metabolites. A notorious example is the ergot kernels, which are sclerotia of the ergot fungus (*Claviceps purpurea*) that grow on rye and other cereals. Ingestion of these kernels causes ergotism, a severe form of food poisoning characterized by symptoms such as hallucinations or vasoconstriction, which can

[9] Calvo (2008).

lead to gangrene and limb loss.[10] Zygomycetes such as *Rhizopus* species, do not form fruiting bodies either. During sexual reproduction, they form the above-mentioned zygospores. Like ascomycete moulds, zygomycetes frequently propagate via asexual spores called sporagiospores (functionally comparable to conidia of ascomycetes). These spores develop on sporangiophores ("spore stalks"), which are visible to the naked eye as dark spots on top of the fluffy white aerial mycelium of *Rhizopus*.

Notably, the different developmental stages—including vegetative mycelium, fruiting bodies, sclerotia and other structures—differ in their gene expression patterns and may, therefore, exhibit variations in nutrient content or toxicological profiles.

2.4 Concurrent Classifications of MMP Under EU Law

MMP intended for placing on the market as food products may be subject to specific regulatory requirements, necessitating compliance with these provisions prior to being placed on the EU market. This may include classification as novel foods, as detailed in the chapter "Novel Food Status of Mycelium and Mushroom Products." In such cases, specific rules regarding pre-market authorisation, conditions of use, and labelling must be adhered to.

MMP can also be used to produce food additives, flavourings, and food enzymes, which are subject to specific regulatory requirements defined at the EU level. MMP may also be classified as processing aids, which are subject to national legislation, as these regulations remain unharmonized within the EU. EFSA guidance documents stipulate that the presence or absence of processing aids in novel food production must be experimentally verified in at least three representative batches of the novel food.[11] If a food enzyme is present in a novel food, its enzymatic activity and any potential residual activity must be reported. Alternatively, if the enzyme is inactivated or removed, the applicant must provide a detailed description of the processes and operational conditions used for its inactivation or removal.

Additionally, MMP may sometimes be marketed as possessing certain properties or bearing health claims, creating ambiguity about whether they should be classified as foods or other product categories, such as medicinal products or food supplements, each governed by distinct regulatory frameworks. The intended use of the substances often serves as the determining factor for their classification.

Furthermore, MMP may be authorised for use in feed or pet food, which are subject to regulatory provisions similar to those for food products. However, this authorization does not imply that such products are suitable for human consumption.

---

[10] Haarmann et al. (2009).

[11] EFSA Panel on Food Contact Materials, Enzymes and Processing Aids (2021).

In this section, we address the definitions that can lead to confusion regarding the classification of MMP as food and outline the corresponding regulatory requirements.

2.4.1 Food Additives

Regulation (EC) No 1333/2008 on food additives (FAR) regulates food additives production and use. They are defined as:

> any substance not normally consumed as a food in itself and not normally used as a characteristic ingredient of food, whether or not it has nutritive value, the intentional addition of which to food for a technological purpose in the manufacture, processing, preparation, treatment, packaging, transport or storage of such food results, or may be reasonably expected to result, in it or its by-products becoming directly or indirectly a component of such foods;[12]

According to the FAR, food additives must, on the basis of available scientific evidence, (1) pose no safety concerns at the level of use proposed; (2) serve a reasonable technological need that cannot be achieved in other ways,[13] considering economic and technological considerations; (3) not mislead the consumer. Moreover, their use must be set at the lowest level necessary to achieve the desired effect.[14] By definition, a food additive is not a food and is added for a technological purpose and not for its nutritional value.

Food additives can be produced using mushrooms and mycelia.[15] To be lawfully placed on the market, these additives must be authorised under Regulation (EC) No 1331/2008 establishing a common authorisation procedure for food additives, food enzymes and food flavourings. Once authorised, food additives are classified into one or more functional classes listed in Annex I of the FAR, where their conditions of use are defined.[16] If there is a significant change in the production process or in the raw materials used, a new authorisation is required before the additive can be placed on the market.[17]

---

[12] Regulation (EC) No 1333/2008, Article 3(2)(a).

[13] Regulation (EC) No 1333/2008, Article 6(2) specifies that all the technological functions that food additives can serve, such as preserving the quality of the product, provide necessary ingredients or constituents for food manufactured for groups with special dietary needs, enhancing or ensuring the maintenance of the organoleptic properties, help in the processing, preparation, treatment, packing, transport or storage of the product.

[14] Regulation (EC) No 1333/2008, Article 11.

[15] The production methods to obtain food additives are not specified in the Annexes of the FAR, so it was not possible to determine how many food additives are currently obtained using mushrooms or mycelia.

[16] Regulation (EC) No 1333/2008, Article 9. The article specifies that "allocating a food additive to a functional class shall not preclude it from being used for several functions", but the classification must be based on the principal technological function.

[17] Regulation (EC) No 1333/2008, Article 12.

Additionally, if food additives, flavourings, or enzymes derived from MMP consist of, contain, or are produced from genetically modified organisms (GMOs), applicants are well-advised to seek authorisation under both the respective sectoral regulation and Regulation (EC) No 1829/2003 on genetically modified food and feed. Several such applications are currently under review.[18]

Annex I of the FAR lists the functional classes in which food additives are divided according to their uses. Classes (1) 'sweeteners',[19] (2) 'colours'[20] and (14) 'flavour enhancers' are particularly interesting for food additives obtained from MMP, such as pigments or bitter blockers. It is important to note that MMP added to foods as nutrients, e.g. to increase the protein content of a drink and/or a snack, and that may be eaten as such are not food additives but simple food ingredients.[21]

Fungi are widely used in the production of food colorants, such as the red and yellow polyketide pigments derived from *Monascus sp*. These pigments are available as pigment extracts or in the traditional form of fermented red rice powder, which has been used in Southeast Asia for over 2000 years. However, these colorants are not approved in the EU due to the potential contamination with citrinin, a toxic metabolite.[22] Using chemotaxonomic rationale, some strains of *Talaromyces* species have been identified as producing *Monascus*-like pigments without toxic metabolites.[23]

As for commercial applications, for example, a Czech company produces natural red colorant produced by fermentation with *Penicillium oxalicum*.[24]

---

[18] Maini Rekdal and et al. (2024).

[19] Regulation (EC) No 1333/2008, Annex I. Sweeteners are defined as "substances used to impart a sweet taste to foods or in table-top sweeteners". Article 7 specifies that a food additive can be classified as sweetener only if it replaces sugars for the production of energy-reduced food, non-cariogenic food or food with no added sugars; or/and if it replaces sugars where this permits an increase in the shelf-life of the food; or/and it is used to produce food intended for particular nutritional uses, meaning *"foodstuffs which, owing to their special composition or manufacturing process, are clearly distinguishable from foodstuffs for normal consumption, which are suitable for their claimed nutritional purposes and which are marketed in such a way as to indicate such suitability."*

[20] Regulation (EC) No 1333/2008, Annex I. Colours are defined as "substances which add or restore colour in a food, and include natural constituents of foods and natural sources which are normally not consumed as foods as such and not normally used as characteristic ingredients of food. Preparations obtained from foods and other edible natural source materials obtained by physical and/or chemical extraction resulting in a selective extraction of the pigments relative to the nutritive or aromatic constituents are colours within the meaning of this Regulation". Article 8 specifies that colours shall restore the original appearance of food of which the colour has been affected by processing, storage, packaging and distribution, whereby visual acceptability may have been impaired; making food more visually appealing; giving colour to food otherwise colourless.

[21] Regulation (EC) No 1333/2008, Article 2(c).

[22] Dufossé et al. (2014).

[23] Dufossé et al. (2014).

[24] Dufossé et al. (2014).

2.4.2 Food Flavourings

Food flavourings are regulated by Regulation (EC) No 1334/2008 on flavourings and certain food ingredients with flavouring properties for use in and on food. Food flavourings are defined as products not intended to be consumed as such, which are added to food in order to impart or modify odour and/or taste.[25] "Flavour enhancers", which are classified as food additives, and "flavourings" have similar but not equal definitions. While "flavourings" are added to impart or modify the flavour of a food, "flavour enhancers" simply increase the existing taste and/or odour and there are defined as: "substances which enhance the existing taste and/or odour of a foodstuff".[26]

Food flavourings also require authorization under Regulation (EC) No 1331/2008. Among the flavorings that can potentially be derived from MMP, natural flavourings are particularly noteworthy, as consumers highly value the naturalness of food products.[27] "Mushroom alcohol", 1-octen-3-ol could be one of such flavourings.[28] Natural flavourings are obtained by physical, enzymatic or microbiological processes from fungi, either in the raw state or after being processed through traditional techniques, including fermentation.[29] The requirements for the labelling of "natural flavouring" are specified Regulation (EC) No 1334/2008.[30]

Many different fermented foods from filamentous fungi have been used to improve sensory characteristics of food, such as soy sauce, miso, tempeh, and katsuobushi, especially in terms of the umami taste.[31]

---

[25] Regulation (EC) No 1334/2008, Article 3(2)(a).

[26] Regulation (EC) No 1333/2008, Annex I.

[27] Siegrist and Hartmann (2020); Monaco et al. (2024).

[28] Hadar and Dosoretz (1991).

[29] Regulation (EC) No 1334/2008, Article 3(2)(c) defines "natural flavourings" as flavourings *"obtained by appropriate physical, enzymatic or microbiological processes from material of vegetable, animal or microbiological origin either in the raw state or after processing for human consumption by one or more of the traditional food preparation processes listed in Annex II. Natural flavouring substances correspond to substances that are naturally present and have been identified in nature"*. Annex II therefore lists all traditional food preparation processes: Chopping, coating, cooling, Heating, cooking, baking, frying (up to 240 °C at atmospheric pressure) and pressure cooking (up to 120 °C), distillation, cutting, drying, emulsification, evaporation, extraction, grinding, infusion, maceration, mixing, peeling, percolation, pressing, refrigeration, pressing, roasting/grilling, squeezing, steeping but also microbiological processes and fermentation. The last two potentially gives the opportunity to derive "natural flavourings" from MMP.

[30] Regulation (EC) No 1334/2008, Article 16.

[31] Diez-Simon et al. (2020).

2.4.3 Food Enzymes

Food enzymes are defined as products obtained from plants, animals or micro-organisms or products thereof, including a product obtained by a fermentation process using micro-organisms, that contain one or more enzymes capable of catalysing a specific biochemical reaction, which are added to food for a technological purpose at any stage of the manufacturing, processing, preparation, treatment, packaging, transport or storage of foods.[32]

Food enzymes may be derived from fungi or their by-products, including substances obtained through fermentation processes involving fungi. Enzymes sourced from fungi can potentially enhance the performance of alternative proteins by modifying their structure to achieve improved texture.

The primary distinction between food and food enzymes lies in their intended use: food enzymes are added to catalyze specific biochemical reactions rather than for their nutritional value.

Currently, the national rules govern the marketing and use of food enzymes, as well as the production of foods using such enzymes. This regulatory framework will remain in place until the Union List of Approved Food Enzymes is established, in accordance with Article 4 of Regulation (EC) No 1332/2008. Currently the list has not been yet created.[33]

2.4.4 Processing Aids

Processing aids are not defined as food additives, food flavourings or food enzymes. Processing aids are substances used to process raw materials. Unlike food additives, processing aids are not present in the final food product, with the exemption of unintentional but technologically unavoidable presence of residues of the substance or its derivatives provided they do not have an adverse effect on human life. If food enzymes act as processing aids, they must be authorised as food enzymes to be placed on the EU market. In jurisdictions in which processing aids require authorisation, some MMP applications have been submitted, for example concerning fungal chitosan derived from *Aspergillus niger*.[34] Regulation No. 606/2009 (later replaced by Commission Delegated Regulation (EU) 2019/934), however, restricted the use of chitosan in wine production as an authorised oenological practice. That regulation specifically authorises certain oenological compounds to be used as a processing aid.[35]

---

[32] Regulation (EC) No 1332/2008, Article 3(2).

[33] European Commission (2024).

[34] FSANZ (2012).

[35] Commission Delegated Regulation (EU) 2019/934, Article 3(1).

2.4.5 Borderline MMP: Between Medicinal Products and Food

The definition of foods and medicinal products are mutually exclusive; however, innovation can bring on the market products that resemble both foods and medicinal properties. Medicinal products are defined as:

(a) Any substance or combination of substances presented as having properties for treating or preventing disease in human beings; or
(b) Any substance or combination of substances which may be used in or administered to human beings either with a view to restoring, correcting or modifying physiological functions by exerting a pharmacological, immunological or metabolic action, or to making a medicinal diagnosis.[36]

Food supplements are defined as

> foodstuffs the purpose of which is to supplement the normal diet and which are concentrated sources of nutrients or other substances with a nutritional or physiological effect, alone or in combination, marketed in dose form, namely forms such as capsules, pastilles, tablets, pills and other similar forms, sachets of powder, ampoules of liquids, drop dispensing bottles, and other similar forms of liquids and powders designed to be taken in measured small unit quantities.[37]

Directive 2002/46/EC establishes a positive list of vitamins and minerals which may be used as ingredients of food supplements. A range of MMP applications are currently being utilized to produce vitamins and other nutrients.[38]

While foods are substances intended to be ingested, for their nutritional content or because of their taste, medicinal products are substances, not necessarily ingested, with the specific objectives of (a) treating or preventing a disease or (b) restoring, correcting, or modifying physiological functions by exerting pharmacological, immunological, or metabolic actions, or which are at least presented in a way that suggest that they could actually have these effects.

Grey areas in the classification of products as foods or medicinal products / food supplements concern products sold as foods, which claim or possess certain nutritional or health properties. Rules applicable to medicinal products would be applicable instead of food law provisions because they concern a different area of regulation. For example, unlike foods, medicinal products must always undergo an authorization procedure and are subject to authorised conditions of use.[39]

It is possible that a product has been historically classified as medicinal product in one Member State of the EU and as food in another.[40] The different classification potentially violates the provisions on the free movement of goods[41] and has been

---

[36] Directive 2001/83/EC, Article 1(2).

[37] Directive 2002/46/EC, Article 2.

[38] Strong et al. (2022).

[39] Directive 2001/83/EC, Article 6 and subsequent articles.

[40] Capelli and Klaus (2009).

[41] Treaty on the Functioning of the European Union, Article 28.

discussed extensively in EU jurisprudence (*Commission vs Germany*,[42] regarding capsules containing garlic powder; Commission vs Spain,[43] on food supplements containing certain herbal medicine). The Court of Justice of the European Union (CJEU) often adopted a two-tiered approach to the differentiation of medicinal products from other products.[44] First, the way a product is presented, including labelling, packaging, forms, and claims, must be taken into account to determine if the average consumer would assume the product has medicinal properties.[45] Second, the actual properties of the product must be assessed.[46] These two conditions cannot be considered separately,[47] and must be read in conjunction with one another.[48] This implies that some sort of effect on human health must be present in a medicinal product.[49] Thus, the presence of active substances and their substantial effects on humans are always critical to determine the final status of products.

2.5 MMP as Agricultural Products

Foods are regularly considered "goods" in the sense of the EU law provisions of the free movement of goods in the internal market.[50] For most cases, MPP can be classified as food. However, they may also classify as "agricultural products," to which specific EU law provisions apply (the Common Agricultural Policy or "CAP").[51] The CJEU has ruled that CAP rules take precedence over internal market provisions for goods, guaranteeing a more favourable treatment to foodstuff which are also agricultural products.[52] In particular, specific rules on competition and on the organisation of agricultural markets may be applicable.[53]

While a general definition of "agricultural products" exists in the Treaties,[54] case-law determined that only the list in Annex I of the Treaty on the Functioning of the EU (TFEU) defines the scope of agricultural products.[55]

[42] *C-319/05* (2007).

[43] *C-88/07* (2009).

[44] *C-140/07* (2009); *C-211/03* (2005); *C-227/82* (1983).

[45] *C-60/89* (1991), Paragraph 23.

[46] *C-140/07* (2009), Paragraph 25.

[47] *C-60/89* (1991), Paragraph 12.

[48] *C-358/13* (2014), Paragraph 37.

[49] Capelli and Klaus (2009); de Boer et al. (2015); Melchor and Timmermans (2009).

[50] Schebesta and Purnhagen (2024), p. 80.

[51] Treaty on the Functioning of the European Union, Title III.

[52] *C-83/78* (1978).

[53] Treaty on the Functioning of the European Union, Article 43.

[54] Treaty on the Functioning of the European Union, Article 38(1) TFEU: "*the products of the soil, of stockfarming and of fisheries and products of first-stage processing directly related to these products*".

[55] C-131/87 (1989).

Mushrooms are not explicitly mentioned in Annex I, but they fall under Chap. 7 of the classification "Edible vegetables and certain roots and tubers." Annex I refers to an old tariff classification system for imported goods that was used internationally before 1976, the Brussel Nomenclature. Council Regulation (EEC) No 2658/87 clarifies the content of each Chapter of the Brussel Nomenclature in relation to the currently used classification system, stating that Annex I Chap. 7 must be read as covering edible mushrooms as well.[56] However, it is disputable whether, mycelia (as a part of fungi) may also be classified as agricultural products similarly to mushrooms.

When it comes to products derived from the processing of mushrooms and mycelia, the "agricultural product" status must be determined on a case-by-case basis. Per Article 38(1) TFEU, agricultural products are the products of the soil, of stock farming and of fisheries and products of first-stage processing directly related to these products. However, not only the results of one single processing action on raw materials are subject to agricultural provisions. What must be considered is the economic interdependence between the basic product and the processed product.[57] When the procession of agricultural raw materials represents only a marginal cost in the production of the endproduct, the end product cannot be considered an agricultural product. For example, mushroom cultivation may fall under the CAP provisions; mycelium biomass produced via fermentation, cut in slices and sold to consumers is likely to be not considered an agricultural product; highly processed end products containing ingredients derived from mushrooms and mycelium e.g. veggie burgers made with protein powders extracted from the mycelium, are likewise not to be classified as agricultural products.

Case Law

C-131/87 - *Commission vs Council* (1989) Court of Justice of the European Union ECLI:EU:C:1989:581.

C-140/07 - Hecht-Pharma (2009) Court of Justice of the European Union ECLI:EU:C:2009:5.

C-185/73, Hauptzollamt Bielefeld v König (1974) Court of Justice of the European Union ECLI:EU:C:1974:61.

C-211/03 - HLH Warenvertrieb and Orthica (2005) Court of Justice of the European Union ECLI: EU: C: 2005: 370.

[56] Regulation (EEC) No 2658/87, Note 2 to Section I – Chapter 7 Council Regulation (EEC) "In headings 0709 to 0712, the word 'vegetables' includes edible mushrooms, truffles…". Heading 07/09 covers other vegetables, fresh or chilled, heading 07/10 vegetables (uncooked or cooked by steaming or boiling in water), frozen, heading 7/11 vegetables provisionally preserve but unsuitable in that state for immediate consumption and heading 7/12 dried vegetables, whole, cut, sliced, broken or in powder, but not further prepared.

[57] *C-185/73* (1974), Paragraphs 12, 13.

C-227/82 - Criminal proceedings against Leendert van Bennekom, (1983) Court of Justice of the European Union ECLI:EU:C:1983:354.
C-319/05 Commission vs Germany (2007) Court of Justice of the European Union ECLI:EU:C:2007:678
C-358/13 - Opinion of Advocate General Bot (2014) ECLI:EU:C:2014:1927.
C-60/89 - Monteil and Samanni (1991) Court of Justice of the European Union ECLI:EU:C:1991:138.
C-83/78 - Pigs Marketing Board v Redmond (1978) Court of Justice of the European Union ECLI:EU:C:1978:214.
C-88/07 - Commission v Spain (2009) Court of Justice of the European Union ECLI:EU:C:2009:123.

Instruments and Legislation

Directive 2001/83/EC on the Community code relating to medicinal products for human use, OJ L 311, p. 67–128.
Directive 2002/46/EC of the European Parliament and of the Council of 10 June 2002 on the approximation of the laws of the Member States relating to food supplements, OJ L 183, p. 51–57.
Directive 2002/46/EC of the European Parliament and of the Council of 10 June 2002 on the approximation of the laws of the Member States relating to food supplements, OJ L 183, 12.7.2002, p. 51–57
Regulation (EC) No 1331/2008 of the European Parliament and of the Council of 16 December 2008 establishing a common authorisation procedure for food additives, food enzymes and food flavourings, OJ L 354, 31.12.2008, p. 1–6
Regulation (EC) No 1332/2008 of the European Parliament and of the Council of 16 December 2008 on food enzymes and amending Council Directive 83/417/EEC, Council Regulation (EC) No 1493/1999, Directive 2000/13/EC, Council Directive 2001/112/EC and Regulation (EC) No 258/97 OJ L 354, p. 7–15.
Regulation (EC) No 1333/2008 of the European Parliament and of the Council of 16 December 2008 on food additives, OJ L 354, 31.12.2008, p. 16–33
Regulation (EC) No 1334/2008 of the European Parliament and of the Council of 16 December 2008 on flavourings and certain food ingredients with flavouring properties for use in and on foods and amending Council Regulation (EEC) No 1601/91, Regulations (EC) No 2232/96 and (EC) No 110/2008 and Directive 2000/13/EC, OJ L 354, 31.12.2008, p. 34–50
Regulation (EC) No 178/2002 of the European Parliament and of the Council of 28 January 2002 laying down the general principles and requirements of food law, establishing the European Food Safety Authority and laying down procedures in matters of food safety, OJ L 31, 1.2.2002, p. 1–24
Regulation (EEC) No 2658/87 of 23 July 1987 of Council of the European Union on the tariff and statistical nomenclature and on the Common Customs Tariff, OJ L 256, 7.9.1987, p. 1–675.

References

Bibliography

Calvo AM (2008) The VeA regulatory system and its role in morphological and chemical development in fungi. Fungal Genet Biol 45:1053

Capelli F, Klaus B (2009) 'Is garlic a food or a drug? How to solve the problem of free movement in the European Union of products that are classified in different ways in the Member States; with specific regards to the delimitation of foodstuffs – including food supplements, novel foods and enriched foodstuffs – and medicinal products. Eur Food Feed Law Rev 4:390

de Boer A, van Hunsel F, Bast A (2015) Adverse food-drug interactions. Regulat Toxicol Pharmacol 73:859

Diez-Simon C and others (2020) Chemical and sensory characteristics of soy sauce: a review. J Agricult Food Chem 68:11612

Dufossé L and others (2014) Filamentous fungi are large-scale producers of pigments and colorants for the food industry. Curr Opin Biotechnol 26:56

Food Safety Authority Australia New Zealand (2012) 'Risk and Technical Assessment Report (at Approval) – Application A1077 Fungal Chitosan as a Processing Aid`, 1, www.https://www.foodstandards.gov.au/

Haarmann T and others (2009) Ergot: from witchcraft to biotechnology. Mol Plant Pathol 10:563

Hadar Y, Dosoretz CG (1991) Mushroom mycelium as a potential source of food flavour. Trends Food Sci Technol 2:214

Kurtzman CP, Fell JW, Boekhout T (2011) Chapter 1 - definition, classification and nomenclature of the yeasts. In: Kurtzman CP, Fell JW, Boekhout T (eds) The yeasts, 5th edn. Elsevier

Londoño-Hernández L and others (2017) Rhizopus Oryzae – ancient microbial resource with importance in modern food industry. Int J Food Microbiol 257:110

Luo H and others (2022) Genes and evolutionary fates of the Amanitin biosynthesis pathway in poisonous mushrooms. Proc Natl Acad Sci 119:e2201113119

Maini Rekdal V and others (2024) Edible mycelium bioengineered for enhanced nutritional value and sensory appeal using a modular synthetic biology toolkit. Nat Commun 15:2099

Melchor SR, Timmermans L (2009) "It's the Dosage, Stupid": The ECJ clarifies the border between medicines and botanical food supplements. Eur Food Feed Law Rev 4:185

Monaco A and others (2024) 'Consumers' perception of novel foods and the impact of heuristics and biases: a systematic review. Appetite 196:107285

Schebesta H, Purnhagen K (2024) EU food law. Oxford University Press

Siegrist M, Hartmann C (2020) Consumer acceptance of novel food technologies. Nat Food 1:343

Spatafora JW and others (2017) The fungal tree of life: from molecular systematics to genome-scale phylogenies. Microbiol Spectrum 5:32

Strong PJ and others (2022) Filamentous fungi for future functional food and feed. Curr Opin Biotechnol 76:102729

Tian F and others (2022) Mycotoxins in Soybean-based foods fermented with filamentous fungi: occurrence and preventive strategies. Compr Rev Food Sci Food Saf 21:5131

Chapter 3
Novel Food Status of Mycelium and Mushroom Product

Abstract This chapter provides a comprehensive analysis of the EU legal framework governing novel foods and pre-market authorization. By examining several examples of consultations regarding the novel food status of mushroom and mycelium-based products (MMPs), it explores the requirements for demonstrating a history of safe use. The chapter then focuses on the technical aspects involved in preparing a novel food application, with particular reference to the most recent guidance documents published by the European Food Safety Authority (EFSA). Furthermore, the chapter critically examines the Qualified Presumption of Safety (QPS) list and other mechanisms designed to streamline the risk assessment process for MMPs. It also highlights several challenges and gaps in toxicological data related to filamentous fungi, providing insights into the complexities of ensuring the safety of these products.

3.1 Scope of the Novel Food Regulation

Per Regulation (EU) No 2015/2283 on novel foods (the Novel Food Regulation or NFR) food business operators must be granted a premarket authorisation before they place a product falling within the definition of a novel food on the EU market.

A food requires authorisation under the NFR if it was not used for human consumption to a significant degree within the EU before 15 May 1997 that is in any state that is now a Member State of the EU. Such food does not benefit from the history of safe use exemption.

The NFR also contains other more specific exemptions: for example, food consisting of, isolated from or produced from animals obtained by traditional breeding practices which have been used for food production within the EU before 15 May 1997 does not constitute a novel food.[1] Also exempt are foods consisting of, isolated

[1] In this case, it is not important to prove the use for human consumption of such food to a significant degree within the EU; however, it is important to prove any use for food production.

A. Molitorisová et al., *Regulation of Mushroom- and Mycelium-Based Products in the EU*, SpringerBriefs in Law, https://doi.org/10.1007/978-3-031-93780-4_3

from or produced from plants or a variety of the same species obtained by traditional propagating practices which have been used for food production within the EU before 15 May 1997, and such food has a history of safe use.[2] Also, a food consisting of, isolated from or produced by a plant or a variety of the same species from non-traditional propagating practices which have not been used for food production within the EU before 15 May 1996, where those practices do not give rise to significant changes in the composition or structure of the food (negatively) affecting its nutritional value, metabolism or level of undesirable substances is not considered a novel food.[3]

Novel foods are either foodstuffs or food ingredients. Food enzymes, food additives, food flavourings and extraction solvents used or intended to be used in the production of foodstuffs or food ingredients do not fall within the scope of the NFR.[4]

3.2 History of Safe Use

Proving history of safe use may result in a food not being considered a novel food, if there is a considerable consumption within the EU before 15 May 1997. The fact that all the individual ingredients of a food product have a history of safe use before 15 May 1997 does not mean that a food would not require authorisation under the NFR.[5]

Submitting information and documents that prove the history of use of the sources as well as the novel food, in case the novel food status was established, may also facilitate safety assessment in the authorisation procedure. For example, if the applicant provides information about normal quantities consumed, this will help assessing whether a new purpose of the novel food would correspond to such quantities or would lead to significantly higher intakes or whether the amounts that are intended to be used would deviate extensively from those normally consumed with common food.

The history of safe use within the EU is not defined in Article 3(2) of the Novel Food Regulation; However, that provision defines history of safe food use in a third country as a case where the safety of the food in question has been confirmed with compositional data and from experience of continued use for at least 25 years in the customary diet of a significant number of people in at least one third country prior to the notification according to the Novel Food Regulation. A use of a food in third countries only, as well as data on non-food use, is not acceptable to demonstrate a

[2] Regulation (EU) No 2015/2283, Article 3(2)(a).

[3] Regulation (EU) No 2015/2283, Article 3(2)(a).

[4] Regulation (EU) No 2015/2283, Article 3(2)(a).

[5] Food Supplements Europe (2019).

history of food use within the EU to avoid the novel food status; however, they could be relevant for assessing the safety of the novel food.[6]

3.2.1 *Extent of Consumption in the EU*

To determine the status of a novel food and for preparing a novel food application, it is essential to provide detailed information about the extent of its consumption, specifying whether this pertains to consumption across the entire EU or within a specific Member State. Details should be provided if the food was consumed only regionally or locally in the EU before 15 May 1997. It should be decided in each individual case whether local use amounts to consumption with significant degree. Other relevant factors to be considered are whether food business operators were present on the market continuously or not or whether the use of the food is linked to local or regional traditions. Information may be also included on whether the food was available as an ingredient for a specific target population, for example for a special medical purpose. That means non-food uses can be listed as well. The following information may be provided to determine the extent of use:

- Quantity of consumption
- Information on the serving size
- The role of the food in the diet (main disk, snack, consumption patter)
- Target population group, etc.[7]

Some of the evidence, such as extensive sales data, provides clear and unequivocal information about the foods' status as novel foods. Individual pieces of information, such as import or distribution lists, cannot provide a sufficient basis for establishing the status of the food as non-novel (Table 3.1).[8]

However, it is more often the case that the evidence that now needs to be presented is more than 25 years old, and, therefore, the whole picture needs to be examined carefully. Comprehensive literature reviews on the history of use and human studies on relevant safety outcomes should be included as well. The applicant may indicate any other supporting information that would assist in determining the novel food status, including information on specific relevant components of the novel food and studies with similar foods with the same or closely related sources or chemical composition. Evidence must be robust, reliable, and based on data taken from trustworthy sources and sources that relate to foods which have been legally on the EU market.[9]

[6] European Commission (2016).

[7] European Commission (2016).

[8] European Commission (2016).

[9] European Commission (2016).

Table 3.1 Evidence in support of history of consumption. Adapted from Human Consumption to a Significant Degree Information and Guidance Document[a]

Type of information	Type of evidence	Possible weighting
Comprehensive sales information	Invoices: Details on the sale of the foodEvidence of large quantities of sale	Very good evidence, if purpose (food use) is indicated
Sales information	Invoices: Details on the sale of the food	Good evidence, if purpose (food use) is indicated
	Catalogues (recognised)Sales brochures	Supporting evidence, if purpose (food use) is indicated
Government import/ export information	Official documents	Supporting evidence, if purpose (food use) is indicated
Expert knowledge	Personal testimonies	Supporting evidence
Supporting information	Magazine articlesRecipe books	Supporting evidence
Other information	Other	Supporting evidence

[a]European Commission (2016)

Information on the search strategy used to retrieve the studies should be indicated, together with sources used to retrieve pertinent data, and terms and limits used (e.g. publication dates, publication types, language, population, default tags).[10] Full study-reports should be provided, if available. Where applicable, the published literature should be reviewed by considering EFSA's systematic review principles.[11]

Also, it should be shown whether the source from which the food is produced is normally consumed as part of a normal diet. Consumption quantities vary depending on the type of the food, for example, herbs versus cereals. They can also vary if one considers a normal diet of the average population in contrast with specific population groups. For a specific population group, information should be provided as to the role of the novel food in the diet, the handling and preparation of the food (fermentation, soaking, peeling, time of cooking, temperature) and on precautions of use. A reference should be made to typical levels of consumption for specific product categories. Any specific processing should be described, including that which aims at reducing adverse effects, such as allergenic or antinutritional substances.[12]

The applicant should also provide information on the availability on the market (distribution via specialised or a limited number of shops or big retail or common supermarkets) and the nature of the selling points (such as door-to-door sale). It is also of interest to assess whether the food has been available on a regular basis or only occasionally or at one time, at a fair trade, for example. If a food had been used at specific occasions like ceremonies, festivities, *et cetera*, such use may be considered significant. However, evidence of commercial availability takes precedence

[10] European Commission (2016).

[11] EFSA (2010a).

[12] Constable (2007).

over evidence of any private domain use. In other words, evidence that mushrooms have only been used as food by some people or that mushrooms have been picked in a forest is only of a limited value. If a food was imported for personal consumption only, it is not a relevant information.[13] Also, a food that had been available on the market only during emergency situations is not considered of "significant use", if it had not been placed on the market also regularly or in a commercially sustainable way. Finally, demonstrating a history of use for other than food uses, such as uses in food supplements or in feed, is also of limited value. A use exclusively in food supplements would also not be considered human consumption to a significant degree. Uses of the food such as in food additives, flavourings, or extraction solvents, cannot be considered as a relevant food use.[14]

In the consultation procedure, Member States may help to determine whether apparently small quantities of a food eaten in the past would render the status of the food not novel under the NFR. For example, it may be that certain mushrooms have been available on local farmers' markets, with limited commercial value, and they may be therefore included in official documentation for edible mushrooms. This could be considered a sufficient proof of significant consumption of those mushrooms. References to the foods in relevant national and EU legislation could also provide good evidence. However, not all EU Member States maintain such lists of edible mushrooms susceptible for commercialisation in their national laws.[15]

The use to a significant degree also does not automatically lead to the exemption from NFR rules if the food in question has been subject to additional processing, such as if this processing alters the composition of the food and is produced from a new source material or by a new production process. In that regard, specific selective extracts of a fungus could be considered a novel food if they have not been used for human consumption as such, even where the source material has been widely consumed. Consideration should also be given to the type of extracts, such as aqueous extract versus other solvent-based extract. By extension, this also applies to other parts of a fungus for which history of safe use cannot be demonstrated, such as fruiting body versus mycelium.

Different parts of a fungus require different assessments of the history of safe use and consequently of safety.[16] This is due to the influence of intrinsic and extrinsic factors on bioactive metabolite profiles, and as such mycelium and fruiting bodies, even of the same fungal species or strain, cannot be regarded as the same product.

The information above can be simplified in the decision-making Table 3.2.

Where food business operators are unsure whether the information in their possession is sufficient to prove that the food concerned has been used for human

---

[13] European Commission (2016).

[14] European Commission (2016).

[15] For example, in Slovakia: Výnos Ministerstva pôdohospodárstva Slovenskej republiky a Ministerstva zdravotníctva Slovenskej republiky zo 7. júna 2000 c. 1541/3/2000 -100, ktorým sa vydáva hlava Potravinového kódexu Slovenskej republiky upravujúca jedlé huby a výrobky z jedlých húb.

[16] Turck et al. (2016).

Table 3.2 Decision making table when establishing history of safe use. Adapted from Human Consumption to a Significant Degree Information and Guidance Document[a]

Mycelium or its source	Yes (evidence)	No (incl. no available information)	Summary
Was consumed by a large number of people as a food or food ingredient throughout in the EU prior to 15 May 1997			
Was consumed only regionally/on a small local scale.			
Was used in the private domain only			
Was available as an ingredient designed for a specific target population			
Was consumed as a normal part of the diet			
Was placed on the EU market and was available for purchase by consumers (e.g. not only in pharmacies, health shops or specific restaurants)			
Has been consumed for a long period of time			
Has been consumed in quantities typical for similar products of the specific food category			
Was harvested from the wild			
Was available as an ingredient for a specific target population group			
Was restricted to individuals with an underlying medical condition.			

[a]European Commission (2016)

3.3 Consultation on Novel Food Status: Procedure

A food business operator may consult the Member State where it first intends to place a food on the market if the food business operator is unsure whether the food falls within the scope of the Novel Food Regulation.[17] This is a voluntary procedure. However, it may be that multiple Member States are being considered for the market launch which means that multiple national authorities may be consulted, and the novel food inventor may choose which to ask. The consulted Member State should reach a conclusion within 4 months (plus 4 months in case of extension).[18] After such a determination, the Commission updates the EU Novel Food Catalogue, a tool developed by the Commission which serves as orientation on whether a product would need an authorisation under the Novel Food Regulation.[19] As such, the Catalogue is a non-exhaustive list of the status of some foods, based on assessments conducted by the competent authorities of Member States.

[17] 'Regulation (EU) No 2015/2283, Article 4.

[18] Commission Implementing Regulation (EU) 2018/456, Article 6(1).

[19] European Commission (2024a).

Fig. 3.1 Adapted from Human Consumption to a Significant Degree Information and Guidance Document [European Commission (2016)]

consumption to a significant degree within the Union before 15 May 1997, they may consult other food business operators or food business operator organizations in order to gather sufficient information.

The decision tree in Fig. 3.1 can be followed to assist applicants with the determination of the novel food status.

Commission Implementing Regulation (EU) No 2018/456 on the procedural steps of the consultation process for determination of novel food status implements Article 4 NFR. The consultation process is initiated by the food business operator with a request which must contain a cover letter, a technical dossier, supporting documentation and an explanatory note clarifying the purpose and relevance of the submitted documentation. The purpose of the technical dossier is to conclude the novel food status. Additional information may be requested for this purpose. Some of the information submitted by the applicant are to be treated as confidential, except of the summary of the studies and where appropriate, the analytical method.[20] According to Annex II that includes a template technical dossier, the applicant should provide description of the food in question, as well as proposed category of the novel food in accordance with Article 3(2)(a) NFR. The applicant should include further characterisation of the food and/or source of the food, where relevant. If this includes organisms, such as fungi, the applicant should specify, where applicable, the part of the organism that was used for human consumption within the Union before 15 May 1997. The applicant should also include information about purity and concentration.[21]

There have been several reasoned statements of national authorities published on the Commission's website, some of which concern mycelium-based and other fungi-based products.[22] Some of them concluded that the foods concerned are considered novel foods since they did not present history of safe and significant consumption in the EU before 15 May 1997, and it was not possible to establish an equivalence relationship between the reproductive and vegetative part of the fungus in terms of nutritional composition, metabolites and potentially bioactive substances. It must be noted that the Novel Food Regulation does not use the term "reasoned statements", nor does it prescribe specific legal effects of the outcome of the consultation process. One way of interpreting the silence of the law is to say that national authorities' statements are not binding on other national authorities or the Commission. However, this would undermine the objective of the consultation process in the regulatory framework characterised by maximum harmonisation. In other words, it would be against the single market principles underpinning EU food law.[23] The Commission also has the power to "overrule" the statement of a Member State about the novel food status by virtue of Article 5 NFR, which prescribes that the Commission may decide whether a particular food falls within the definition of novel food. In case another Member State does not agree with the determination of the novel food status as performed by the first Member State, it may ask the Commission to proceed with the determination of the novel food status following Article 5. Also, since national authorities inform one another in the consultation process and gather necessary information for the novel food status assessment, it is

---

[20] Commission Implementing Regulation (EU) 2018/456, Article 9.

[21] Commission Implementing Regulation (EU) 2018/456, Annex II.

[22] European Commission (2023).

[23] Van der Meulen (2013).

unlikely that their opinions would differ from one another. This happened in 2022 when two national authorities—Finnish and Swedish were simultaneously asked about the status of mycelium powder of *Rhizopus oligosporus*, confirming its non-novel food status.

3.4 Consultation on Novel Food Status: Examples

Out of the information available on the Commission's website, it is possible to discern different regulatory practices of national authorities when it comes to the reasoning about the novel food status. Each reasoned statement enriches the collective practice and may help food innovators with the determination of their food products in the development pipeline. Regarding the history of safe use, the Finnish authority concluded in November 2024 that chaga, a conk produced by the fungus *Inonotus obliquus* on its host tree (birch), is a novel food. This was despite the evidence that the conk of *Inonotus obliquus* mushroom was used in the tea/coffee preparations in the nineteenth century and during the wartime. The use of the food has however not continued until recently and cannot be considered significant. Also, the use during emergencies does not constitute evidence of significant use.[24]

As for the mycelium powder of *Rhizopus oligosporus*, in June 2022, the Finnish authority concluded that *Rhizopus oligosporus* had been traditionally produced and consumed as part of tempeh, a traditional food from Indonesia.[25] The Finnish Food Authority thereby suggested that tempeh 'has been widely consumed in the EU already before 15 May 1997'.[26] Although, the nutritional composition of the novel food under consideration differed from tempeh, the authority took the absence of mycotoxins in the novel end product for a proof of the fact that the production process of the mycelium powder did not give rise to significant changes in composition as compared to tempeh. The Swedish authority in its statement in June 2022 simply reiterated that fresh form of the food (tempeh) did not differ from the dried form (powder) which is therefore not a novel food.[27] To what degree tempeh was consumed in the territory of today's EU before 1997, neither the Finnish nor the Swedish authority attempted to answer. 'Significant degree' of human consumption is a relative term, and the consumption of the food in question in third countries does not constitute relevant evidence of the consumption within the Union. The Finnish authority considered 'tempeh' traditional food from Indonesia which is itself subject to the notification or authorisation process according to Section II of the Novel Food Regulation. There is no separate Union list of notified or authorised

[24] Finnish Food Authority (2024).

[25] Finnish Food Authority (2022a).

[26] Finnish Food Authority (2022a).

[27] Livsmedelsverket (2022).

traditional foods; such products are simply put on the Union list of novel foods.[28] Tempeh is not listed as a 'traditional food'; that does not mean that traditional tempeh fermentation using soybeans has not a long history of use, however, *stricto sensu* should not be considered traditional food under the Novel Food Regulation. At the same time, it must be recalled that *Rhizopus oligosporus* is only one type of filamentous fungal strain used in tempeh fermentation in Indonesia. Others can be *Aspergillus* or *Mucor*, and these strains could also be considered non-novel for fermentation purposes following the reasoning of the Finnish and Swedish authorities provided that the absence of mycotoxins in the final product is established.[29] On the other hand, in 2011, the Commission authorised the placing on the market of fermented black bean extract (Touchi extract), i.e. a protein-rich powder obtained by water extraction of small soybeans fermented with *Aspergillus oryzae*, as a novel food ingredient for use in food supplements.[30] In this case, it was the specific food being significantly different from existing foods, which was not used in the EU before 15 May 1997, that triggered the novel food status, not the source material or the fermentation.[31] This extract is typically produced from small soybeans grown in the Sichuan province of China. *A. oryzae* is a well-established fungus employed in the production of soy sauce, sake and miso, and the processing uses conventional processing techniques.

The Swedish and Finnish decisions were issued following an earlier decision of the Irish Food Safety Authority in 2022 that dried yeast fermentate produced by *Saccharomyces cerevisiae* is not novel because fermentation by *Saccharomyces cerevisiae* is a well-known process in food production in the EU.[32] The Irish authority concluded that yeast fermentate did not significantly differ from existing non-novel fermented foods that also contain yeast biomass (dead or alive), fermentation by-products and fermentation medium.[33] The same yeast was used in the process of fermentation of wheat germ extract which was subject to consultation by the Hungarian authority in 2020. The Hungarian National Food Chain Safety Office, however, concluded that the product as such showed no evidence of a history of

---

[28] Commission Implementing Regulation (EU) 2017/2470 of 20 December 2017 establishing the Union list of novel foods in accordance with 'Regulation (EU) No 2015/2283 of the European Parliament and of the Council on novel foods, C/2017/8878, OJ L 351, 30.12.2017, pp. 72–201.

[29] Tian et al. (2022), p. 5131.

[30] Commission Implementing Decision 2011/497/EU.

[31] Commission Implementing Decision 2011/497/EU.

[32] Food Safety Authority of Ireland (2022a). By the reasoning presented in this statement, one could infer that fermentation with *Neurospora intermedia* or *Fusarium venenatum* could also escape the definition of novel food in the EU, although this has not been confirmed in the Novel Food Catalogue or by any other authority.

[33] Commission Implementing Decision 2011/762/EU. In 2009, the competent authorities of Ireland were also dealing with the request 'to place yeast beta-glucans on the market as a novel food ingredient for use in a variety of foods, including beverages, for the general population as well as in food supplements and in foods for particular nutritional uses with the exception of infant formulae and follow-on formulae.'

consumption to a significant degree.[34] Here, the reasoning was different from the one of the Irish Food Safety Authority since the Hungarian authority did not establish the history of consumption of *Triticum aestivum* which is a common wheat, nor did it note that fermentation by *Saccharomyces cerevisiae* is a well-known process similar to the Irish authority. In 2022, the Irish Food Safety Authority also evaluated another request for consultation for torula yeast (*Cyberlindnera jadinii*) used in biomass fermentation.[35] With the evidence of history of consumption from the Czech Republic (however, without mentioning the degree thereof), the authority concluded a non-novel status of the product. The evidence of consumption of mushrooms made be drawn from national lists of edible mushrooms that may be introduced to the market, for example in the Slovak Food Codex.[36]

In another case,[37] in 2019, involving fungal biomass (*Aspergillus oryzae*) grown on a medium enriched with minerals, the Irish Food Safety Authority referred to a Danish list of food cultures in support of the conclusion that *Aspergillus oryzae*'s history of consumption in the EU precedes 15 May 1997.[38] This outcome of the Irish consultation process concerning mycelium differs from the opinion of the Spanish authority on *Lentinula edodes* dehydrated mycelium powder. In the latter statement, the Spanish authority concluded that the product which is derived from a mushroom commonly named *Shiitake* is a novel food despite a history of safe and significant consumption in the EU before 15 May 1997 of the reproductive part of the fungi.[39] However, the Spanish authority concluded that such evidence of consumption is irrelevant as it cannot be used to demonstrate that also mycelium, i.e. the vegetative part of the fungi, had been consumed in the EU before the relevant date. Moreover, the Spanish authority consistently opined in a series of consultations (dehydrated mycelium powder of *Agaricus blazei,*[40] *Coprinus comatus,*[41] *Pleurotus ostreatus,*[42] *Pleurotus eryngii,*[43] *Polyporus umbellatus,*[44] *Grifola frondose,*[45] *Hericium erinaceus,*[46] and *Ganoderma lucidum*[47]) that equivalence between the reproductive part and the vegetative part of the fungi cannot be established in terms of nutritional

[34] National Food Chain Safety Office of Hungary (2020).

[35] Food Safety Authority of Ireland (2022b).

[36] Slovak Food Codex (2000) https://www.nahuby.sk/clanok.php?clanok_id=19.

[37] Food Safety Authority of Ireland (2019).

[38] Barzee et al. (2021). In fact, *Aspergillus oryzae* have been utilised for at least 3000 years in the production of koji and soy sauce.

[39] Spanish Agency for Food Safety and Nutrition (2019a).

[40] Spanish Agency for Food Safety and Nutrition (2019b).

[41] Spanish Agency for Food Safety and Nutrition (2019c).

[42] Spanish Agency for Food Safety and Nutrition (2019d).

[43] Spanish Agency for Food Safety and Nutrition (2019e).

[44] Spanish Agency for Food Safety and Nutrition (2019f).

[45] Spanish Agency for Food Safety and Nutrition (2019g).

[46] Spanish Agency for Food Safety and Nutrition (2019h).

[47] Spanish Agency for Food Safety and Nutrition (2019i).

composition, metabolites, and active substances. The same conclusion regarding the mycelium of *Lentinula edodes* was reached by the Belgian nationality authority in the consultation concerning a plant protein concentrate.[48] The Belgian authority referred simply to the lack of evidence of human consumption to a significant degree within the EU before 15 May 1997 specifically of mycelium but not mushroom of the fungi. The authority concluded that mycelium, in this context, is to be regard as an ingredient of the product—it is a food consisting of, isolated from or produced from microorganisms, fungi or algae. In this case, the novel food status of the other ingredients and of the process applied has not been evaluated.[49]

These statements concerning *Lentinula edodes* were preceded by a 2007 request to the UK authorities to place a mycelium extract from *Lentinula edodes* on the market as a novel food ingredient.[50] Also, the German authorities confirmed the novel food status for blackcurrants fermented with mycelium of *Wolfiporia cocos* in 2021 because fermenting with the mycelium of *Wolfiporia cocos* is a novel production process in the EU.[51]

Two other fungi products were identified as novel foods in 2019: *Agaricus bisporus* grown on selenium-enriched[52] and on vitamin B12 substrates[53] was considered a novel food by the UK Food Standards Agency due to new production processes although the fungi had a significant history of consumption in the EU prior 15 May 1997. This was, however, not confirmed by the Irish Food Safety Authority in December 2022 after the previous UK decision was reversed in June 2022.[54] The UK authorities obtained additional evidence of a history of consumption of 'button mushrooms' in the UK and the EU with varied and raised levels of B12 and selenium resulting from farming practices that supplemented the growing substrate.[55] The UK statement did not specify which EU Member States were included in the evidence. It must be noted that the evidence of consumption in the UK is currently irrelevant for the determination of the novel food status in the EU, perhaps except for Northern Ireland.[56] The Irish authority was more specific in stating its reasons: for example, it

[48] Belgium Federal Public Service of Health, Food Chain Safety and Environment (2019).

[49] Belgium Federal Public Service of Health, Food Chain Safety and Environment (2019). The production technology was based on submerged cultivation in sterilised liquid medium. Lentinula edodes mycelium was cultivated in a liquid aerobic fermentation process. The mycelium of the cultivated Lentinula edodes is submerged in defined medium, comprising glucose, malt extract, soy peptone and yeast extract. Controlled fermentation conditions (temperature, aeration rate, pH) are applied. The biomass is removed by filtration and the resulting fermentation liquid is the raw material for the lentian-based products. Final concentrations of lentinan are adjusted by dilution with water. The liquid is sterilised by head and sodium benzoate is added as a preservative. Based on this opinion, company MycoTechnology, Inc. submitted a dossier in the novel food authorisation process for pea and rice protein fermented by Shiitake mycelia.

[50] Commission Decision 2011/73/EU.

[51] Bundesamt für Verbraucherschutz und Lebensmittelsicherheit (2021).

[52] Food Standards Agency (2019).

[53] Food Standards Agency (2019).

[54] Food Safety Authority of Ireland (2023).

[55] Food Standards Agency (2022).

[56] Protocol 2019/C 384 I/01. Article 5(4).

cited a decision of the Finnish government to supplement compound fertilisers used for cereal and grass production with selenium. It is however unclear how this may serve as evidence of 'consumption' and how it corresponds to the Commission's guidance document on demonstrating 'human consumption to a significant degree'.[57]

National courts have the possibility (and in case of the court of the highest instance also an obligation) to pose a reference for a preliminary ruling to the CJEU to the effect of clarifying a provision of EU law. With regards to novel foods, the *Landesgericht für Zivilrechtssachen Graz* in Austria lodged a question in February 2022 whether 'sprouted buckwheat flour with a high spermidine content' is a novel food, inasmuch as only sprouted buckwheat flour without a raised spermidine content was used for human consumption to a significant degree within the EU before 15 May 1997 or has a history of safe food use thereafter, irrespective of how the spermidine comes to be in the sprouted buckwheat flour.[58] One needs to wait for the answer, however, the preliminary reference raises further important interpretive concerns as regards the term 'production process' in the novel food status under the EU Novel Food Regulation.[59] In January 2023, AG Campos Sánchez-Bordona issued his opinion in the matter. He pointed out that in June 2022, Austria notified the Commission with its reasoned statement that the food in question is a novel food, citing 'possible significant changes to the composition' induced by the altered production process.[60] The Austrian authority also pointed out that the documentation submitted with the request did not prove that the food in question was consumed to a significant degree within the EU before 15 May 1997. AG Campos Sánchez-Bordona suggested that the concept of 'history of safe food use in a third country', which entails compositional data and continued use for at least 25 years in the customary diet of a significant number of people in at least one third country, can be transposed to the concept of 'history of safe use within the Union'.[61] By that interpretation, such conditions (compositional data + at least 25 years of consumption + continued use + significant number of people) must be satisfied cumulatively. The Court endorsed these views of AG Campos Sánchez-Bordona.[62]

In 2022 and early 2023, two more consultation procedures were concluded: one concerning the fungal biomass of *Paecilomyces variotii* assessed by the Finish

[57] European Commission (2016).

[58] C-141/22 (2023).

[59] Molitorisová et al. (2021).

[60] Federal Ministry of Social Affairs, Health, Care and Consumer Protection (2022).

[61] *C-141/22 Opinion of Advocate General Campos Sánchez-Bordona* (2023).

[62] On 24 May 2023, the Court (Seventh Chamber) ruled that "a food, such as sprouted buckwheat flour with a high spermidine content, which was not used for human consumption to a significant degree within the European Union before 15 May 1997, constitutes a 'novel food' within the meaning of that provision given that, first, it is obtained from a plant, secondly, it is not apparent that its safety has been confirmed with compositional data and from experience of continued use for at least 25 years in the customary diet of a significant number of people in at least one country of the European Union, and, thirdly and in any event, it is not obtained by propagating practices, within the meaning of that provision."

authority which concluded that the protein ingredient is to be considered a novel food.[63] In January 2023, the Spanish Agency for Food Safety and Nutrition confirmed the novel food status for edible biomass produced by adding mycelium of *Pleurotus ostreatus* to a plant substrate.[64] The Spanish authority concluded that there was not history of consumption to significant degree of fungal mycelium of *Pleurotus ostreatus*. This decision came after the Spanish authority already evaluated the novel food status for dehydrated mycelium powder of the same fungal strain, as noted above.

In 2024, there were three additional relevant consultations: the Czech Agriculture and Food Inspection Authority concluded that a cell wall of the yeast *Saccharomyces cerevisiae* in powder form is not a novel food.[65] However, the same species of yeast enriched with iron is considered to be a novel food.[66] It was also confirmed that mycoprotein derived from *Fusarium venenatum* A 3/5 is not a novel food.[67]

Two more reasoned statements concerning fermentation processes may be of interest: In the case concerning fermented apricot kernel cream, evaluated by the Austrian national authority, it was noted that its ingredients are not novel (such as apricot kernels) and are used for similar purposes in products available on the EU market (such as almond drink, almond yoghurt). The production process of fermented apricot kernel cream just includes common and widely used production practices (process technologies) and can be compared to existing products such as nut drinks and nut yoghurts. Cultures used for the fermentations are also used frequently in other products on the market, while the driver of fermentation is added sugar. The common and traditional use of apricot kernels is well demonstrated and documented in various legislation (flavouring regulation, import control on pesticide residues and contaminants), as well as in the Codex Alimentarius Austriacus and the German Food Code (*Lebensmittel- und Bedarfsgegenstände- und Futtermittelgesetzbuch*).[68] On the other hand, a history of consumption to a significant degree within the EU prior to 15 May 1997 has not been demonstrated for fermented wheat germ extract.[69]

3.4.1 Authorisations and Decisions Pursuant to Regulation (EC) 258/97

Furthermore, several Member States and the Commission issued decisions concerning the placing on the market of fungal products prior to the coming into effect of the Novel Food Regulation. In 2009, for example, the Dutch Ministry of Health,

---

[63] Finnish Food Authority (2022b).

[64] Spanish Agency for Food Safety and Nutrition (2023).

[65] Czech Agriculture and Food Inspection Authority (2024a).

[66] Czech Agriculture and Food Inspection Authority (2024b).

[67] Czech Agriculture and Food Inspection Authority (2024c).

[68] Austrian Federal Ministry of Social Affairs, Health, Care and Consumer Protection (2021).

[69] National Food Chain Safety Office of Hungary (2020).

Welfare and Sport allowed arachidonic acid-rich oil from the fungus *Mortierella alpina* to be placed on the market after conducting an initial assessment in accordance with Article 6 of Regulation (EC) 258/97.[70] Another fungi product, lycopene from *Blakeslea trispora* (in other words a carotenoid from mould), was assessed by the UK authorities in 2007, whereas the EFSA issued its opinion in 2009.[71] Other authorisations are mentioned in the respective references in this contribution. The established Union list of novel foods includes the novel foods authorised or notified under Article 4, 5 or 7 of Regulation (EC) No 258/97, including any existing authorisation conditions.

3.4.2 EFSA's Opinions

When a national authority establishes that a food is a novel food, a company may decide to file a dossier to the Commission to apply for a marketing authorisation. After that, the Commission may request an opinion from the EFSA in which the EFSA assesses the food's safety. Based on the application, the EFSA must adopt its opinion within nine months of receipt of the request from the Commission, however, this process can be extended as a result of "suspended timeline" or "stop the clock" procedures, where the regulator requests additional information or data from the applicant in order to assist their assessment.

Issuing EFSA's opinions on safety of various fungi-derived products as requested by the Commission already presupposes that a product is categorised as novel food. In 2022, the EFSA evaluated safety of freeze-dried mycelia of *Antrodia camphorata* intended to be marketed as food supplements.[72] Also in 2022, the EFSA completed and issued a positive evaluation of vitamin D2 mushroom powder (*Agaricus bisporus*),[73] while it already found that vitamin D2 mushroom powder (*Agaricus bisporus*) is safe in 2021[74] and 2020[75]—these applications were submitted by different applicants who were awarded the protection of proprietary data in accordance with Article 26 of the Novel Food Regulation. All assessed vitamin D2 mushroom powders were UV-light treated. Such assessments are also interesting from the perspective that UV-treated mushrooms (but not powder) with increased vitamin D

[70] Ministry of Health, Welfare and Sport of the Netherlands (2016). The product was later authorised by Commission Decision of 12 December 2008 authorising the placing on the market of arachidonic acid-rich oil from *Mortierella alpina* as a novel food ingredient under Regulation (EC) No 258/97 of the European Parliament and of the Council (notified under document number C(2008) 8080). The product was first requested to be placed on the market in 1999 in the Netherlands.

[71] Commission Decision 2009/365/EC.

[72] EFSA Panel on Nutrition, Novel Foods and Food Allergens (2022a).

[73] EFSA Panel on Nutrition, Novel Foods and Food Allergens (2022b).

[74] EFSA Panel on Nutrition, Novel Foods and Food Allergens (2021a).

[75] EFSA Panel on Nutrition, Novel Foods and Food Allergens' (2020a).

content could also be placed on the on the market in the EU pursuant to Article 4(2) of Regulation (EC) No. 258/97 after an initial assessment performed by the Food Safety Authority of Ireland in 2016[76] and 2017 for *Agaricus bisporus*[77] and Sweden in 2017.[78] All of these authorisations originate from different applicants. Also, *Saccharomyces cerevisiae* (baker's yeast) was UV-treated and risk-evaluated in 2014[79] and evaluated for safety of its extended use in foods in 2021.[80] *Saccharomyces cerevisiae* is a species of yeast included in the QPS list. It is interesting to note that in April 2022, the Italian Ministry of Health led consultation into the novel food status of whey treated with UV radiation and concluded that the food is not novel because UV treatment did not result in significant changes in the composition or structure of the food.[81] The EFSA also issued two opinions on the safety of *Yarrowia lipolytica*—a common yeast added to the QPS list—once in 2019 with regard to a dried and heat-killed biomass[82] and in 2020 with regard to its application in chromium-enriched biomass—both with positive outcome.[83]

The EFSA's opinion on pea and rice protein fermented by shiitake *(Lentinula edodes;* synonym: *Lentinus edodes)* mycelia is of interest.[84] The EFSA's Panel on Nutrition, Novel Food and Food Allergens concluded that no toxicological studies or human data are required on the novel food. The Panel dropped the previous requirement on the toxicological studies for two reasons: the nature of the novel food and the history of safe use, including its generally recognised as safe (GRAS) status in the US. This is noteworthy because the GRAS status should have in principle no legal relevance in the EU, but it underlines an apparent willingness from EU regulators to take into consideration the assessment decisions and expertise of international interlocutors. This may potentially mark a novel development in EU regulatory behaviours.

Furthermore, the EFSA's Panel differentiated between the history of use of the source and history of the novel food itself. As regards the history of the source, the Panel noted that the mycelium fermentation agent (*Lentinula edodes*) was not used

---

[76] Food Safety Authority of Ireland (2016). The decision of the Food Safety Authority of Ireland does not specify which fungi species the application concerned.

[77] Food Safety Authority of Ireland (2017).

[78] Commission Implementing Decision (EU) 2017/2355 of 14 December 2017 authorising the placing on the market of UV-treated mushrooms as a novel food under Regulation (EC) No 258/97 of the European Parliament and of the Council (notified under document C(2017) 8474), C/2017/8474, OJ L 336, 16.12.2017, pp. 52–54.

[79] EFSA Panel on Dietetic Products, Nutrition and Allergies (2014).

[80] EFSA Panel on Nutrition, Novel Foods, Food Allergens (2021b).

[81] Italian Ministry of Health (2022). However, the authority also noted the possibility to use the UV radiation for the purposes presented by the company is ruled at national level, meaning that, the use of the treatment with UV radiation must be authorized by the Member State where the business operator wants to market the food.

[82] EFSA Panel on Nutrition, Novel Foods and Food Allergens (2019).

[83] EFSA Panel on Nutrition, Novel Foods and Food Allergens (2020b).

[84] EFSA Panel on Nutrition, Novel Foods and Food Allergens (2022c).

for food production in the EU prior to 15 May 1997, although the fruiting body has been consumed "all over Europe long before". It is unclear whether the Panel took the history of the source or the history of novel food, or both, as relevant for the toxicological assessment. The Panel also referred to the EFSA's opinion from 2010 in which it assessed the safety of *Lentinus edodes* extract, which, however, only referred to an established history of consumption of the fungi's fruiting body.[85] This contrasts with a previous EFSA's evaluation. In 2010, the EFSA also delivered a scientific opinion on the safety of chitin-glucan which is derived from a fermentation process of mycelium of *Aspergillus niger*.[86] Despite noting that *Aspergillus niger* is a non-toxic and non-pathogenic strain with a history of safe use, the applicants conducted and delivered toxicological assessments.

It is to be seen how future applications for novel food authorisations involving fungal mycelium will be treated by the EFSA, and how novel food status of mycelium products will be determined by national authorities in consultation procedures. Currently, an application for the authorisation of fungi protein from *Fusarium flavolapis* filed in 2021 awaits a scientific assessment.[87] Also, in 2024, an application for mycelium from oyster mushrooms produced via solid-state fermentation of *Pleurotus ostreatus* was submitted.[88]

3.5 Novel Food Application

The novel food application must comply with the Administrative Guidance for the Preparation of Applications on Novel Foods pursuant to Article 10 NFR which was most recently revised by EFSA in 2024.[89]

Food business operators have a general obligation to verify whether the food which they intend to place on the EU market falls within the scope of the NFR. This entails verifying whether it can be classified as foods that are excluded from the NFR, such as food enzymes, and whether the food was used to a significant degree before 15 May 1997 within the EU. Second, where no grounds for general exclusion from the application of the NFR are given, a food business operator needs to assess whether the food falls within one of the categories defined in the NFR. For carrying out such verification, the product's composition, characteristics, source material and production process need to be assessed at minimum.[90] Also, a fungus that is used to produce a novel food but is not part of the final novel food product to be assessed by EFSA does not fall under the NFR. The safety of the fungus in the context of the

[85] EFSA (2010b).

[86] EFSA (2010c).

[87] European Commission (2021).

[88] European Commission (2024b).

[89] EFSA Panel on Nutrition, Novel Foods and Food Allergens (2024a).

[90] C-383/07 (2009); Food Supplements Europe (2019).

manufacturing process of the novel food is considered during the risk assessment of the novel food. If the fungi are ingested, however, their novel food status needs to be verified.[91]

Where a food business operator has carried out a verification of a new food, it is recommended that the food business operator prepares a written and reasoned record of this verification and retains this on file in case the status of the food is questioned by national control authorities. This record should contain the arguments and evidence justifying the classification of the food as not falling under the scope of the NFR.

If a food business operator has determined that its product is a novel food, prior to entering the market in the EU the food business operator must apply for authorisation of the novel food to the Commission via an online application form. There are no application fees, although there are substantial costs associated with the dossier creation.

The application for the authorisation of a novel food should be preferably in English and must contain (Table 3.3):

- A cover letter drafted in accordance with a template[92]
- Administrative data
- A technical dossier containing administrative data and scientific data
- A summary of the dossier (summary data).[93]

The dossier submitted should enable a comprehensive risk assessment of the novel food.

Novel foods must be tested as they are intended to be marketed. To this purpose, the applicant must provide documentation on the procedure and strategy followed when gathering all relevant data.[94]

On receipt of an application, the Commission verifies whether the application falls within the scope of the NFR and whether it fulfils all the prescribed requirements. The Commission informs the applicant, EU Member States and the EFSA about the validity of the application. The Commission requests EFSA's opinion not later than one month after having verified its validity.[95]

The EFSA assesses an application for a novel food, usually following a request (mandate) from the European Commission. The EFSA analyses the nutritional, toxicological and allergenic properties of the novel food and its production process and delivers an opinion on the safety of the novel food.[96] The EFSA then publishes third party comments and conducts a public consultation to collect new or additional scientific evidence or data that may be beneficial to enabling the EFSA to

[91] Food Supplements Europe (2019).

[92] Commission Implementing Regulation (EU) 2017/2469, Annex I.

[93] EFSA Panel on Nutrition, Novel Foods and Food Allergens (2024a).

[94] EFSA Panel on Nutrition, Novel Foods and Food Allergens (2024a).

[95] Regulation (EU) 2015/2283, Article 11.

[96] EFSA Panel on Nutrition, Novel Foods and Food Allergens (2021c).

assess the safety of the application. More general comments about stakeholder views on the desirability of the novel food entering the EU market are not considered in scope during the consultation process. The outcome of the public consultation is an annex to the EFSA's opinion.

An applicant may ask for a confidentiality request covering certain information that the applicant submitted in the dossier.[97] The non-confidential version of the application dossier can be subject to a public consultation.

An application may be also submitted from producers to modify the conditions of use, the specifications, additional specific labelling requirements and post-marketing requirements of a novel food product which has already been authorised by the EFSA.

In the online application form, the applicant must first provide information according to the categories under which the novel food falls.[98]

3.6 General Conditions for Placing on the Market of Novel Foods

The Commission authorises and includes a novel food in the Union list if (a) the food does not, on the basis of the scientific evidence available, pose a safety risk to human health; (b) the food's intended use does not mislead the consumer, especially when the food is intended to replace another food and there is a significant change in the nutritional value; (c) where the food is intended to replace another food, it

---

[97] The following information may be considered confidential by the EFSA: (a) The manufacturing or production process, including the method and innovate aspects thereof, as well as other technical and industrial specifications inherent to that process or method, except for information which is relevant to the assessment of safety; (b) Commercial links between a producer or importer and the applicant where applicable; (c) Commercial information revealing sourcing, market shares or business strategy of the applicant; (d) Quantitative composition of the subject matter of the request, except for information which is relevant to the assessment of safety, (e) Where applicable, information provided in detailed description of starting substances and starting preparations and on how they are used to manufacture the novel food, and detailed information on the nature and composition of the specific foods or food categories in which the applicant indents to use that novel food, except for information which is relevant to the assessment of safety; (f) Where applicable, detailed analytical information on the variability and stability of individual production batches, except for information which is relevant to the assessment of safety; (g) Any other personal data except for the name and address of the application, the names of authors of published or publicly available studies supporting such requests and the names of all participants and observers in meetings of the Scientific Committee and the Scientific Panels, their working groups and any ad hoc group meeting on the subject matter; and (h) Personal data of individuals involved in testing on vertebrate studies or in obtaining toxicological information.

[98] Such as class 12.9—protein products, excluding products covered in category 1.8, i.e. dairy analogues, including beverage whiteners. This is not meant to reflect the regulatory categories that appear in Administrative Data section, as outlined in Article 3(2) of the Novel Food Regulation.

does not differ from that food in such a way that its normal consumption would be nutritionally disadvantageous for the consumer.

3.7 Notification of Studies

The Regulation (EU) 2019/1381[99] requires notification of all studies presented in the dossier for authorisation submitted to the EFSA. The notification must be done before the study begins. Business operators must notify the title and scope of any study commissioned and carried out, as well as the laboratory or testing facility carrying out that study, and its starting and planned completion dates. The notification obligation also applies to laboratories and testing facilities in the Union, as well as to those in third countries if provided in relevant agreements and arrangements.

In case of failure to notify studies, an application for authorisation is to be considered invalid, unless the applicant provides a valid justification. If the studies were notified but were not included in the application, the authorisation procedure is also suspended unless the applicant provides a valid justification, or the applicant includes all scientific studies. The same is the case if the applicant withdraws a notification of a study.

On request, the raw data for the individual studies, published and unpublished, undertaken by the applicant, or on their behalf, must be provided.[100]

One analysis of the Commission's decisions and available novel foods' dossiers revealed that the majority of applications were rejected due to a lack of or delay in the notification of the scientific studies supporting the application and, only in a few cases, due to a lack of data available to the EFSA for risk assessment purposes.[101]

[99] Regulation (EU) 2019/1381.

[100] EFSA Panel on Nutrition, Novel Foods and Food Allergens (2024a).

[101] Dongo and Della Penna (2023).

3.8 Technical Dossier Checklist

Table 3.3 Technical dossier checklist supporting a novel food application

Pre-application information	Information on studies that have been notified in EFSA's database of study notifications, but not submitted in the application or withdrawn.
Identity of the novel food	Identity and description of the novel food.
Production process	Detailed description of the production process, specific production process steps, including measures of production control, and quality and safety assurance. A non-confidential summary of the production process, including all steps of the process with a general description of the operational conditions and safety-related parameters. Flow chart of production process.
Compositional data	Characterisation of the novel food. Qualitative andquantitative data on the composition. Information on impurities/contaminants and stability of the novel food. Information on analytical methods.
Specifications	Key parameters characterising the novel food. Rationale for the selection, limits and information on the method used for each parameter.
History of use of the novel food and/or of its source	Comprehensive literature review on history of use of the novel food and its source, and human studies on relevant safety outcomes. Information on the search strategy used to retrieve the studies.
Proposed uses and use levels and anticipated intake of the novel food	Target population. Form of uses, food categories, maximum amounts as consumed. Anticipated and combined intake of the novel food from other sources. Exposure to undesirable substances and precautions and restrictions of use.
ADME	Data on absorption, distribution, metabolism and excretion in human and animals, considering nutritional and toxicological impact of the novel food and following a tiered testing approach. Quantification of relative bioavailability or demonstration of bioavailability for novel foods proposed as new nutrient sources.
Toxicological information	Toxicological studies, genotoxicity studies, repeated-dose toxicological studies. Rationale for the toxicity testing strategy applied. Human data.
Nutritional information	Intake of nutrients, antinutritional factors, specific considerations for novel foods proposed as new sources of micronutrients and for novel protein sources.
Allergenicity	Protein content of the novel food. Information onallergenic proteins in the novel food if derived from known allergenic foods. Comprehensive literature review on information on the novel food, its source and used raw materials, when the allergenic potential is unknown.
Concluding remarks	Overall considerations of the information that supports the safety of the novel food under the proposed conditions of use. Consider toxicological relevant components, toxicity studies, adverse effects and sources of uncertainties. For novel foods that are also nutrient sources, conclusions on both safety and (relative) bioavailability.
References	
List of annexes and checklist	

Source: Adapted from Administrative guidance for the preparation of novel food applications in the context of Article 10 of Regulation (EU) 2015/2283 [EFSA Panel on Nutrition, Novel Foods and Food Allergens (2024a)]

3.9 Novel Food Categories and Identification

Second, when filing a novel food application, it is necessary for applicants to designate the most appropriate food category under which the food in question falls in accordance with Article 3(2) of the Novel Food Regulation. Categories listed under Article 3(2) are not exclusive categories, that means more than one category can be chosen. The proposed classification is based on the chemistry, production process and source of novel foods. The most appropriate food categories for novel food candidates of mushrooms, mushroom-derived products and fungal mycelium are likely to be:

(i) food consisting of, isolated from or produced from fungi.
This category may include:

- traditional fermentation, using novel fungi to process ingredients derived from plants or other sources, which can result in products with unique flavours and textures.[102] Where fungi in novel food is capable of multiplication, they are considered active agents. Where fungi are used in the manufacturing process to produce novel foods, they are referred to as production strains.[103] There are over 20 types of fermented foods using filamentous fungi.[104]
- biomass fermentation using high protein content of fungi to produce large quantities of protein. Biomass is food made from fungi in which cells are not necessarily capable of multiplication, but their genetic material may still be detected. Biomasses are complex mixtures where not all constituents can be fully characterised or identified. For those, comprehensive qualitative and quantitative data for the classes of naturally or chemically derived components should be submitted. If the applicant anticipates that a new protein will be created during the production process, s/he must submit a characterisation of the protein profile.
- fungi used to produce specific compounds that are natural metabolites of those fungi, where the compound is harvested, concentrated, or further purified. For example, chitin-glucan from *Aspergillus niger* was authorised to be placed on the EU market as a novel food ingredient to be used in food supplements.[105] The Commission also authorised the placing on the market of lycopene obtained by extraction and crystallisation from a fungal fermentation of *Blakeslea trispora*.[106]

(ii) food consisting of, isolated from or produced from cell culture or tissue culture derived from fungi. This category was included in the list of novel food categories after cultivated meat became familiar to the regulators.[107]

---

[102] Boukid et al. (2023).

[103] EFSA Panel on Nutrition, Novel Foods and Food Allergens (2024a).

[104] Wikandari et al. (2023).

[105] Commission Decision 2011/76/EU.

[106] Commission Decision 2006/721/EC.

[107] European Parliament (2015).

In both categories, the word 'fungi' covers mushrooms' fruiting bodies, fungal mycelium or food that consists or is produced from fungi cell culture or tissue culture.

Two further classifications may be considered:

(iii) food with a new or intentionally modified molecular structure, where that structure was not used as, or in, a food within the Union before 15 May 1997.

For example, it may be that genetically modified fungi strains could be used in fermentation processes but are not further detectable in the product and thus ingested.[108] That means that only food products free from DNA and cells from the production microorganism may be considered novel foods, otherwise they are subjected to additional regulatory assessments and risk evaluations by the EFSA.[109] If a genetically modified fungus is used as a production strain, its purpose, characterisation and structure must be provided.[110] The presence of such production strain must be analysed. In a recent case, the EFSA evaluated a novel food whose primary constituent is a structurally identical 3-fucosyllactose, a naturally occurring trisaccharide occurring in human milk, produced through fermentation by a genetically modified microorganism (*Escherichia coli*) and subsequent purification.[111]

(iv) food resulting from a production process not used for food production within the Union before 15 May 1997, which gives rise to significant changes in the composition or structure of a food, affecting its nutritional value, metabolism or level of undesirable substances. In this case, the classification of a novel food depends on the specific manufacturing/use practice. In a fungi application, "processing" may include the process of fermentation, picking or mushroom cultivation.

A representative decision tree can look as shown in Fig. 3.2.

Scientific requirements for the taxonomic identification and hazard assessment of a fungal strain depend on its role, whether it is genetically modified and in the case of yeast, whether it is included in the Qualified Presumption of Safety list. In case a fungus serves as an active agent or biomass, the following scientific information is required:

(1) Unambiguous taxonomic identification at species level according to internationally recognised database or culture collection with the status of International Depository Authority
(2) Certification of deposition (including accession number)
(3) For yeasts, the species and strain identity according to internationally accepted methods should be verified

[108] Ronchetti et al. (2024).

[109] EFSA Scientific Committee (2022).

[110] EFSA Scientific Committee (2022).

[111] EFSA Panel on Nutrition, Novel Foods and Food Allergens (2020c).

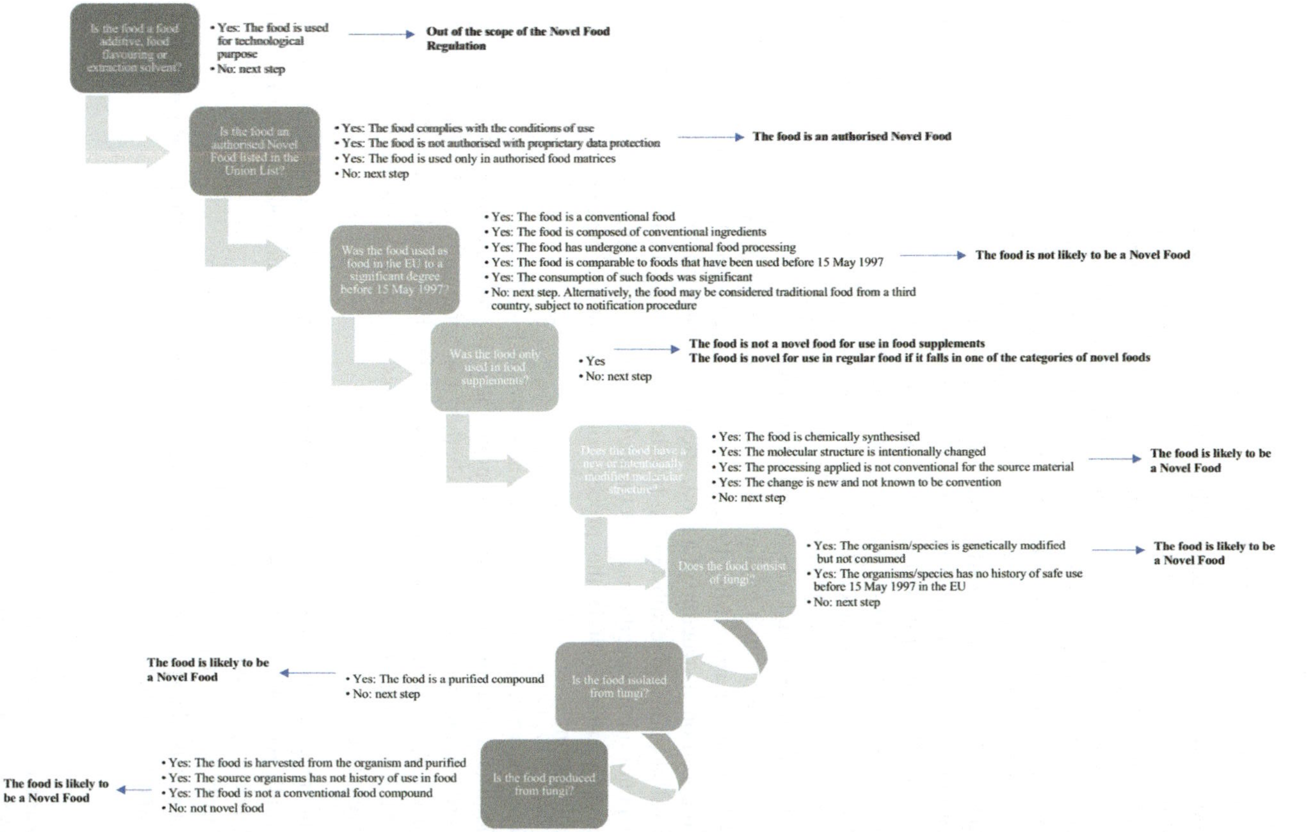

Fig. 3.2 A representative decision tree. Adapted from the Guidance for food business operators on the verification of the status of a new food [Adapted from Food Supplements Europe (2019)]

(4) Characterisation of genes of potential concern, i.e. toxigenicity and pathogenicity traits
(5) Assessment of the capacity of the microbial strain to produce antimicrobials of clinical relevance
(6) Whole genome sequence data according to the most up-to-date versions of the available and applicable EFSA scientific outputs
(7) Presence of viable cells for biomasses. The respective concentration of viable cells and non-viable cells should be reported.[112]

In case of macroscopic fungi, or their parts, the following information must be provided for each source:

(1) Scientific (Latin) name and taxonomy (family, genus, species, and if applicable subspecies, variety with author's name, chemotype, strain) according to the international codes of nomenclature for macroscopic fungi, accepted synonyms and trivial or common names used to identify the novel food intended to be marketed

    - The International Code of Nomenclature for algae, fungi, and plants is the set of rules and recommendations that govern the scientific naming of all organisms traditionally treated as algae, fungi, or plants, whether fossil or non-fossil, including blue-green algae (Cyanobacteria), chytrids, oomycetes, slime moulds, and photosynthetic protists with their taxonomically related non-photosynthetic groups (but excluding Microsporidia). Provisions of the Code relating solely to names of fungi are gathered in its Chapter F. The content of Chapter F may be modified by action of the Fungal Nomenclature Session of an International Mycological Congress (IMC). Chapter F was revised by the closing plenary session of the 11th IMC, held in San Juan, Puerto Rico. The resulting San Juan Chapter F was published as an open-access article.[113]

(2) Verification of the identity according to internationally recognised databases and methodology and, if available deposition in an internationally recognised culture collection with access number

    - FUSARIOID-ID database will bring together key generic and species descriptions, protocols and identification methodologies for working with the genus *Fusarium* and other fusarioid sister genera.[114]
    - MycoBank is an on-line database documenting mycological nomenclatural novelties (new names and combinations) and associated data.[115]
    - CABI Compendium on invasive fungal species.[116]

[112] EFSA Panel on Nutrition, Novel Foods and Food Allergens (2024a).
[113] May et al. (2019).
[114] See https://www.fusarium.org.
[115] See https://www.mycobank.org/.
[116] See https://www.cabidigitallibrary.org/journal/cabicompendium/about.

(3) Growing conditions to produce the source organism (i.e. cultivated or from the wild, conditions of cultivation, growing region of source organisms and season of harvesting).[117]

If the novel food consists of, is isolated from or produced from cell culture or tissue culture derived from fungi, the applicant should state the following information, where available:

(1) Identity of the source (taxonomic information on family, genus, species, subspecies, variety) according to the international codes of nomenclature
(2) Verification of the identity of a fungus according to internationally recognised databases and methodology
(3) Laboratory or culture collection sourced
(4) Information on the identity of cells: parts of the fungi sourced, cell type isolated, genetic and phenotypic identity and stability of cell lines.[118]

3.10 Whole Genome Sequencing of MMP

In October 2021, the EFSA published an updated Scientific Guidance for the submission of dossiers on food enzymes which is supposed to assist applicants in the preparation and presentation of dossiers for the safety evaluation of food enzymes,[119] including food enzymes that are of microbial origin. It includes filamentous fungi and yeasts in this category.

To identify and comprehensively characterise these microorganisms for risk assessment, whole genome sequence (WGS) analysis is required for the bacterial and yeast strains used as production strains and recommended for filamentous fungi. According to the EFSA, "WGS data provide information for the characterisation of the strains regarding their functional traits of concern (e.g. virulence factors, production of or resistance to antimicrobials of clinical relevance of clinical relevance, production of known toxic metabolites)". In the meanwhile, the EFSA prepared a follow-up document to support applicants in the preparation of WGS data for the characterisation of microorganisms used in food chain.[120] In this document, the EFSA summarizes technical recommendations for genome sequencing, relevant quality criteria, as well as essential information to be reported. Since sequencing techniques are evolving steadily, these recommendations are anticipated to undergo regular updates and revisions. For instance, each microorganism should be cultivated before DNA extraction as a pure culture. Total DNA extradition methods should include

[117] EFSA Panel on Nutrition, Novel Foods and Food Allergens (2024a).

[118] EFSA Panel on Nutrition, Novel Foods and Food Allergens (2024a).

[119] EFSA Panel on Food Contact Materials, Enzymes and Processing Aids (2021).

[120] EFSA (2024a).

both chromosomal and extra-chromosomal elements (e.g. plasmids) and need to be described in detail.

Different sequencing strategies can be employed. The sequencing reads may be assembled *de novo* and annotated or mapped to a reference genome/database. In any case, long-read sequencing technologies, or a combination of short-read and long-read sequencing are essential for bacterial strains and are also highly recommended for yeasts and filamentous fungi. If *de novo* assembly is conducted, assembler software, version and parameters need to be indicated. Thresholds for base quality (e.g. per-base PHRED score of at least 20 for short-reads) and sequencing coverage (>30 fold) are to be indicated. The number of contigs, i.e., contiguous, reconstructed pieces of the original DNA sequence obtained in an assembly, should not exceed 1000. If a higher number of contigs is produced (resulting in a highly fragmented genome assembly), a justification should be provided.

To check for completeness of the genome sequence, the number of highly conserved genes, such as Benchmarking Universal Single-Copy Orthologs (BUSCO) genes, present in the assembly should be reported. This parameter indicates the completeness and quality of the assembly.[121] Ideally, >90% complete matches to BUSCO gene set from the most closely related group of yeasts/filamentous fungi should be present in the assembly.

When WGS data is available, taxonomic identification should be conducted by phylogenomic analysis. Phylogenetic trees can be constructed by comparing conserved sequences, such as those targeted in the Assembling the Fungal Tree of Life (AFToL) project, including internally transcribed spacer (ITS) regions. These sequences are concatenated and used as a basis for inferring phylogenies in comparison with related genomes. Furthermore, alignment of the whole genome to a complete reference genome from the same species can be performed, follow by computation of the average nucleotide identity (ANI). For ANI-based taxonomic identification, a similarity of at least 99% is recommended when the strain under assessment is compared to the type strain. The methodology and sequence(s) used for comparison, as well as relevant sequence identities should be reported. The nomenclature and taxonomy of fungi, including yeasts, is covered by the International Code of Nomenclature for algae, fungi and plants (ICN). Applicants are referred to the website Mycobank.[122]

Whole-genome sequencing (WGS) data should then be analysed to detect genes of concern, including those associated with virulence, resistance to clinically significant antimicrobials, or the biosynthesis of harmful metabolites. *De novo* assembled sequences can be examined using a search or comparison-based approach against curated databases. Identified matches should be summarized. For each match, database sequence name and accession number, the function of the encoded protein, sequence identity, and the percentage of the subject sequence length covered should be reported.

---

[121] See https://busco.ezlab.org/.

[122] See https://www.mycobank.org/.

If the strain is genetically modified according to the definition in Directive 2001/18/EC, the genetic modification should be described. The characterisation of the structure of the genetic modification should be done using WGS data for yeasts and is recommended for filamentous fungi. Sequences and methods used for sequencing and comparison should be described. All genetic modifications, both intended and unintended, must be reported based on the alignment between the genetically modified microorganism (GMM) and the reference strain. The alignments between the GMM and the reference strain must be included. A map (graphical representation) should indicate all insertions, deletions, and substitutions identified in the genome. For each affected open reading frame (ORF), the amino acid sequence and the proposed function and metabolic role must be provided. For filamentous fungi for which WGS is not available, all the steps to obtain the genetic modification should be described in order to identify all genetic material potentially introduced into the recipient/parental microorganism.

As for food enzymes, filamentous fungi and yeast, the guidance includes detailed requirements on microbiological purity. Where possible presence of compounds of known toxicity (e.g. mycotoxins) arising from the fermentation is indicated by literature searches or WGS analysis of the production strain, the applicant should determine the concentration of these compounds in the food enzyme. Filamentous fungi and yeast should not exceed 100 CFU/g in the food enzyme measured according to the prescribed standard.

3.11 Uncertain Taxonomic Position

The uncertainty of the taxonomic position of an organism is certainly a serious concern. Only if the identity of an organism is unequivocal, meaningful safety assessments can be made. Notably, the metabolic profile of fungi significantly differs not only between species (e.g. between the button mushroom or the death cap) but by also vary between subspecies, varieties, strains or isolates of the same species. These terms typically refer to stable variations within the same species, characterized by distinct genetic and phenotypic traits. They could, for instance, share a high degree of genomic sequence identity but differ in subtle morphological traits or, importantly, in bioactive compound profiles. The reason behind this diversity is genetic variation within species. Unfortunately, relatively minute genetic alterations like point mutations or small deletions and insertions into genes or regulatory genetic elements might result in significant phenotypic changes such as the activation of inactivation of toxin production. An instructive example is the "kōji" mold, *Aspergillus oryzae*, which is used in Japan since centuries for fermentation of soybeans to produce sake, soy sauce or miso paste. Kōji molds have a long history of safe use and *A. oryzae* was granted GRAS status in the US.[123] Notably, however,

[123] Daba et al. (2021).

gene sequencing of *A. oryzae* revealed that it is actually a very close relative of *A. flavus*, an infamous aflatoxin producer and it was suggested that *A. oryzae* should be regarded as a variety within the *A. flavus* species.[124] However, *A. oryzae* is still treated as a separate species, mainly for practical reasons and food safety considerations. There are several genetically distinct isolates of the kōji mold, however, most of them display genetic variations in their aflatoxin gene clusters compared to *A. flavus* and *A. parasiticus*.[125,126] These variations range from deletion of 40% of the gene cluster to subtle changes that leave large portions of the cluster intact. It is hypothesized that genetic mutations, that led to a loss of toxin production, have occurred during the "domestication" of the natural occurring *A. flavus* into atoxigenic *A. oryzae.* During its prolonged propagation under fermentation conditions, i.e. under the absence of the necessity to produce toxins as a fitness factor, mutations in the complex aflatoxin production machinery were allowed to accumulate.[127] However, it remains worrisome that some isolates of *A. oryzae*, despite its domestication, still contain near-intact aflatoxin clusters, as well as other biosynthetic genes for the production of compounds known to exert effects on human health.[128,129]

Given the significant metabolic variation even among closely related isolates, ensuring the precise taxonomic classification of a filamentous fungus is essential for its safe use. For microorganisms including fungi, taxonomic identification down to the species level should be performed and the strain level or even isolate level should be named if possible. Unfortunately, taxonomic identification of fungi is challenging and often requires expert knowledge. Depending on the case, is requires a combination of morphological, phylogenomic, and chemotaxonomic methods. A good example for the difficulty and importance of taxonomic certainty is again the discrimination of *A. flavus*, the aflatoxin producer, and *A. oryzae*, the koji mold used in food fermentation. Their genomes display over 99.5% average nucleotide identity, hampering simple genetic approaches to discriminate the two taxonomic groups.[130] Therefore, classical morphological criteria such as spore morphology remained essential as a means for the identification of isolates.[131] Phylogenomic approaches based on whole-genome sequencing might offer a solution to this issue but require detailed analyses of single-nucleotide variations.[132] Therefore, the phylogenomic identification of a new isolate as *A. oryzae* should not substitute for a comprehensive assessment of its potential to produce aflatoxins and other mycotoxins.

---

[124] Kurtzman et al. (1986).

[125] Chang (2019); Tominaga et al. (2006).

[126] Chang (2019); Tominaga et al. (2006).

[127] Chang (2019); Tominaga et al. (2006).

[128] Jørgensen (2007); Tominaga et al. (2006).

[129] Jørgensen (2007); Tominaga et al. (2006).

[130] Chang (2019).

[131] Jorgensen (2007).

[132] Chang (2019).

It is also noteworthy that the taxonomy of filamentous molds is complicated by a range of factors. This is illustrated by the fact that different reproductive stages of the same fungal species were traditionally given different names. For example, the asexual reproductive stage of a fungal species, the anamorph, often bore a different name than the sexual reproductive stage, the teleomorph. The genus name *Aspergillus* refers to the anamorph stage only, while the teleomorph of *Aspergillus nidulans* is named *Emericella nidulans*. Adding to the potential for confusion, some fungi produce multiple morphologically distinct anamorphs, while others, like most species of *Aspergillus*, lack a known teleomorph and were traditionally referred to as *fungi imperfecti* (Deuteromycota). With the adoption of the *one fungus = one name* principle in fungal nomenclature, the practice of assigning separate names to anamorph and teleomorph stages was largely discontinued.[133] However, legacy names are still commonly encountered in scientific literature and industry. Furthermore, the taxonomic placement of certain isolates, as well as the naming of taxonomic units (TUs), is under constant revision by the scientific community. As a consequence of name changes, the number of synonyms for a single species is constantly increasing, further complicating safety assessments based on previous reports. These examples might suffice demonstrate why the uncertainty of the taxonomic unit is a serious concern raised by the EFSA and should not be underestimated by food producers.

3.12 Specifications

Novel food specifications provided by applicants are key parameters (chemical, physicochemical, nutritional and microbiological) that characterise and substantiate the identity and safety of the novel food. The applicant must provide a rationale for the selected parameters. The parameters should be presented in a table format with minimum and/or maximum specification limits and include:

- Proximate analytes (protein, lipids, carbohydrates, ash and moisture).
- The major groups of constituents within the novel food (amino acids, proteins, carbohydrates, lipids, inorganic ions, polyphenols, alkaloids, terpenes).
- Other characteristic constituents (carotenoids, polyphenols, terpenes, number of viable/non-viable fungi, etc.).
- Safety-relevant parameters (toxins, heavy metals, pathogens, impurities or degradation products, etc., such as heavy metals, mycotoxins, pesticides). For substances produced by microbial fermentation, the presence of metabolites of safety concern should be investigated. Any substances of concern should be analysed.[134]

[133] Hawksworth et al. (2011).

[134] EFSA Panel on Nutrition, Novel Foods and Food Allergens (2024a).

  - For example, production involving *Fusarium venenatum* which enjoys history of safe use and is available on the EU market may result in cytotoxic trichothecene metabolites, such as 4,15-diacetoxyscirpenol.[135]
  - Fermentation involving *Aspergillus niger* may also result in formation of fumonisins.[136]
  - Currently, there are around 100 novel food specifications regarding mycotoxins in 23 novel foods, mainly aflatoxins, ochratoxin A and deoxynivalenol.[137]

- Quality- and stability-related parameters that may impact safety (microbial hygiene indicators, water activity, etc.).[138]
- Employed analytical methods and their sensitivity.
  - For example, the Kjeldahl method for the determination of nitrogenous constituents in *Lentinula edodes* mycelium extract or the Karl Fisher method for the determination of water in tomato lycopene oleoresin.
  - More specifically described methods include the indication of the chromatographic conditions and the type of column in case of alpha-cyclodextrin or trehalose.
  - One may also refer to a specific method published in a scientific journal, such as in case of the determination of nattokinase activity in fermented soya bean extract using the method described by Takaoka et el.
  - 2 novel foods that contain GMOs also lay down specifications demonstrating their absence: ice structuring protein type III HPLC 12 and L-alanyglutamine.
  - The current specification of microbiological parameters does not refer to standardised methods even though they are decisive for the detection or quantification of a microorganism.[139]

Minimum parameters must be included in case of nutrients present in the novel food intended to replace another food in the diet (i.e. be an alternative of a conventional food product). This is especially relevant in case of alternative proteins because they are (often but not always) intended as substitute products for conventionally derived protein-based products.

A comprehensive literature search should be performed to retrieve published compositional data for the source and parts used in the novel food, as well as compositional aspects linked to the production process.

Specifications should be demonstrated by certified analyses, with full description and references, on preferably at least five representative independently produced (non-consecutive) batches of the novel food for each proposed production process. Analyses should be preferably performed by accredited facilities. Representative means that the samples have the same size, containers, conditions, etc. (see Table 3.4).

---

[135] Dufossé et al. (2014); Wiebe (2002).

[136] Frisvad (2007).

[137] López Sánchez de León et al. (2023).

[138] Dufossé et al. (2014).

[139] López Sánchez de León et al. (2023).

Table 3.4 Specifications of novel foods involving fungi as retrieved from Commission Implementing Regulation (EU) 2017/2470

Antrodia camphorata mycelia powder	**Description/Definition:** The novel food is the freeze-dried mycelia of the fungus *Antrodia camphorata* (strain BCRC 39106), which has been grown by solid-state cultivation. The freeze-dried mycelia are then milled into a powder. *Antrodia camphorata* is a synonym of *Taiwanofungus camphoratus* (family: *Fomitopsidaceae*). **Characteristics/Composition:** Loss on drying (Moisture): <10% Carbohydrates: ≤ 80 g/100 g Protein: ≤ 20 g/100 g Ash: ≤ 6 g/100 g Fat: ≤ 6 g/100 g Total triterpenoids: 1.0–10.0 g/100 g Antroquinonol: 1.0–20.0 mg/g **Heavy metals:** Arsenic: < 0.5 mg/kg **Microbiological criteria:** Total aerobic microbial count: $\leq 10^3$ *CFU/g Total yeast and mould count: ≤100 CFU/g *Escherichia coli*: Not detected in 10 g *Salmonella* spp.: Not detected in 25 g *Staphylococcus aureus*: Not detected in 10 g *CFU: Colony Forming Units
Arachidonic acid-rich oil from the fungus *Mortierella alpina*	**Description/Definition:** The clear yellow arachidonic acid-rich oil is obtained by fermentation of the non-genetically modified strains IS-4, I49-N18, FJRK-MA01 and CBS 210.32 of the fungus *Mortierella alpina* using a suitable liquid. The oil is then extracted from the biomass and purified. Arachidonic acid: ≥40% by weight of the total fatty acid content Free fatty acids: ≤0.45% of the total fatty acid content Trans fatty acids: ≤0.5% of the total fatty acid content Unsaponifiable matter: ≤1.5% Peroxide value (PV): ≤5 meq/kg Anisidin value: ≤20 Acid value: ≤1.0 KOH/g Moisture: ≤0.5%

Chitin-glucan complex from *Fomes fomentarius*	**Description/Definition:** Chitin-glucan complex is obtained from the cell walls of the fruit bodies of the fungus *Fomes fomentarius*. It consists primarily of two polysaccharides: Chitin, composed of repeating units of N-acetyl-D-glucosamine (CAS No: 1398-61-4); β-(1,3)(1,6)- D-glucan, composed of repeating units of D-glucose (CAS No: 9041-22-9). The production process consists of several steps, including: cleaning, reduction in size and grinding, softening in water and heating in an alkaline solution, washing, drying. No hydrolysis is applied during the production process. Appearance: Powder, odourless, flavourless, brown **Purity:** Moisture: ≤15% Ash: ≤ 3.0% Chitin-glucan: ≥90% Ratio of chitin to glucan: 70:20 Total carbohydrates, excluding glucans: ≤0.1% Proteins: ≤2.0% Lipids: ≤ 1.0% Melanins: ≤8.3% Additives: None pH: 6.7–7.5 **Heavy metals:** Lead (ppm): ≤1.00 Cadmium (ppm): ≤1.00 Mercury (ppm): ≤0.03 Arsenic (ppm): ≤0.20 **Microbiological criteria:** Total mesophilic bacteria: ≤.10^3/g Yeast and moulds: ≤10^3/g Coliforms at 30 °C: ≤10^3/g *E. coli*: ≤10/g *Salmonella* and other pathogenic bacteria: Absence/25 g

(continued)

Table 3.4 (continued)

Chitosan extract from fungi (*Agaricus bisporus*; *Aspergillus niger*)	**Description/Definition:** The chitosan extract (containing mainly poly(D-glucosamine)) is obtained from stems of *Agaricus bisporus* or from the mycelium of *Aspergillus niger*. The patented production process consists of several steps, including: extraction and deacetylation (hydrolysis) in alkaline medium, solubilisation in acidic medium, precipitation in alkaline medium, washing and drying. Synonym: Poly(D-glucosamine) Chitosan CAS number: 9012-76-4 Chitosan formula: $(C_6H_{11}NO_4)_n$ Appearance: fine free-flowing powder Aspect: Off –white to slightly brownish Odour: Odourless **Purity:** Chitosan content (% w/w dry weight):≥85 Glucan content (% w/w dry weight): ≤15 Loss on drying (% w/w dry weight): ≤10 Viscosity (1% in 1% acetic acid): 1–15 Degree of acetylation (in % mol/wet weight): 0–30 Viscosity (1% in 1% acetic acid) (mPa.s): 1–14 for chitosan from Aspergillus niger; 12–25 for chitin from *Agaricus bisporus* Ash (% w/w dry weight): ≤3.0 Proteins (% w/w dry weight): ≤2.0 Particle size: >100 nm Tapped density (g/cm^3): 0.7–1.0 Fat binding capacity 800 × (w/w wet weight): pass **Heavy metals:** Mercury (ppm): ≤0.1 Lead (ppm): ≤1.0 Arsenic (ppm): ≤1.0 Cadmium (ppm): ≤0.5 **Microbiological criteria:** Aerobic count (CFU/g): $\leq 10^3$ Yeast and mould count (CFU/g): $\leq 10^3$ *Escherichia coli* (CFU/g): ≤10 Enterobacteriaceae (CFU/g): ≤10 *Salmonella*: Absence/25 g *Listeria monocytogenes*: Absence/25 g

Mycelial extract from Shiitake mushroom (*Lentinula edodes*)	**Description/Definition:** The novel food ingredient is a sterile aqueous extract obtained from the mycelium of *Lentinula edodes* cultivated in a submerged fermentation. It is a light brown, slightly turbid liquid. Lentinan is a Lentinan is a β-(1-3)-d-glucan with β-(1-6) branches which has a molecular weight of approximately 5×10^5 Daltons, a degree of branching of 2/5 and a triple helical tertiary structure. **Purity/Composition of the mycelial extract from *Lentinula edodes*:** Moisture: 98% Dry matter: 2% Free glucose: <20 mg/ml Total protein: < 0.1 mg/ml N-containing constituents: <10 mg/ml Lentinan: 0.8–1.2 mg/ml\ Bradford method Kjeldahl method
Pea and rice protein fermented by *Lentinula edodes* (Shiitake mushroom) mycelia	**Description:** The novel food is produced from the fermentation of a mixture of 65 % pea and 35 % rice protein concentrates by the mycelia of the Shiitake mushroom (*Lentinula edodes*) followed by heat treatment to terminate the fermentation and a series of drying steps to form a powder. **Characteristics/Composition:** Protein (% dry weight, N × 6.25): ≥75.0 Moisture: ≤7.0 Total fat (% dry weight): ≤ 10.0 Ash (% dry weight): ≤ 10.0 Carbohydrates (% by calculation): ≤ 15.0

(continued)

Table 3.4 (continued)

	Mycotoxins: Aflatoxin B1 (µg/kg): < 1.0 Aflatoxin B2 (µg/kg): < 1.0 Aflatoxin G1 (µg/kg): < 1.0 Aflatoxin G2 (µg/kg): < 1.0 Aflatoxin total (B1+B2+G1+G2) (µg/kg): < 3.0 Heavy metals: Arsenic (µg/g): < 0.1 Cadmium (µg/g): < 0.1 Lead (µg/g): < 0.3 Mercury (µg/g): < 0.1 **Microbiological criteria:** Total aerobic microbial count: < 1000 CFU/g (CFU: Colony Forming Units) Total yeast/moulds count: < 100 CFU/g Coliforms: ≤ 10 CFU/g *Salmonella* spp.: Absent in 25 g *Escherichia coli*: < 10 CFU/g *Listeria monocytogenes*: Absent in 25 g
UV-treated mushrooms (*Agaricus bisporus*)	**Description/Definition** Commercially grown *Agaricus bisporus* to which UV light treatment is applied to harvested mushrooms. UV radiation: a process of radiation in ultraviolet light within the wavelength of 200-800 nm. Vitamin D_2 Chemical name: (3*β*,5*Z*,7*E*,22*E*)-9,10-secoergosta-5,7,10(19),22-tetraen-3-ol Synonym: Ergocalciferol CAS No: 50-14-6 Molecular weight: 396,65 g/mol **Contents** Vitamin D_2 in the final product: 5–20 µg/100 g fresh weight at the expiration of shelf life.

Vitamin D2 mushroom powder	**Description/Definition** Vitamin D_2 mushroom powder is a granular powder made from homogenised *Agaricus bisporus* mushrooms that have been exposed to UV light. The mushrooms are washed, homogenised and suspended in water to produce a mushroom slurry. The mushroom slurry is passed under a UV lamp. The slurry is then filtered, dried and ground, producing vitamin D_2mushroom powder. UV radiation: A process of radiation in ultraviolet light within a range of wavelength similar to those UV-treated novel foods authorised under the novel food regulation. **Characteristics/Composition** Vitamin D_2 content: 1000–1300 μg/g of mushroom powder Moisture: ≤ 10.0% Ash: ≤ 13.5% Heavy Metals Lead (as Pb): ≤ 0.5 mg/kg Cadmium: ≤ 0.5 mg/kg Mercury: ≤ 0.1 mg/kg Arsenic: ≤ 0.3 mg/kg **Mycotoxins** Aflatoxins (sum of B1+B2+G1+G2): <4 μg/kg **Microbiological criteria:** Total plate count: ≤ 5000 CFU/g Yeast and mould: ≤ 100 CFU/g *Salmonella* sp.: Absent in 25 g *Staphylococcus aureus*: ≤ 10 CFU/g *Escherichia coli*: ≤ 10 CFU/g Coliforms: ≤ 10 CFU/g *Enterobacteriaceae*: ≤ 10 CFU/g *Listeria monocytogenes*: Absent in 25 g

In the absence of some parameters, it is difficult to identify the novel food, for example, as is the case of the absence of purity specification of some extracts. It has been also noted that it would be more useful to refer to the criteria and conditions that analytical methods of contaminants of novel foods should meet rather than to specific analytical methods, since, as the Spanish food safety regulator AESAN points out "…the progress in analytical techniques means that the specific methods established may become obsolete over time."[140]

3.13 Production Process

New production techniques may trigger the novel food status and subsequent assessment requirements even where the organism, or its part, used in the food has a demonstrable history of safe use itself. From the market perspective, one of the most promising considerations is the production of protein biomass or whole cut "meats" as meat substitutes using mycelium in fermentation processes. Whether in the production of whole cut alternative "meats" using filamentous fungi strains or in the production of biomass with other fungi strains, fermentation can play a crucial role in the production of these innovative products.

Many fermentation processes use traditional techniques using intact live microorganisms to modulate and process plant-derived ingredients.[141] Traditional techniques can be said to have a demonstrated history of safe use, and therefore do not raise any novelty considerations or specific or unique safety issues. Most conventional fermentation processes using microorganisms with a history of use for such purposes would not be covered by the Novel Food Regulation as they would not raise novelty issues. However, mycelium of different fungi strains often represents a novel fermentation agent. It may also appear, although rarely, in co-cultures with other traditionally used fermentation agents, such as yeast or lactic acid. Companies which are planning to introduce a novel food using mycelium as a fermentation agent may use different types of fermentation processes. Among these, one may find solid-state fermentation, air fermentation or submerged fermentation, or fermentation in continuous fermentation tanks, fed-batch or in batches. Some companies experiment with recombinant or precision fermentation.[142] There are different reasons why companies opt for different types of fermentation for different fungi strains, such as the growth time, growth space, or control of physicochemical parameters. Fermentation with mycelium may lead to the formation of filamentous and pellet mycelial biomass without the formation of fruiting bodies.

The employment of new technologies in the production process resulting in different production processes can have different food safety implications from the

[140] López Sánchez de León et al. (2023).

[141] Good Food Institute (2024).

[142] Ho (2021).

perspective of food safety assessors. For that matter, the production process should be described in detail in the novel food application to provide enough information that will serve as the basis for the evaluation of the bioavailability, nutritional value, and safety. The applicant should inform in this respect whether a production process is novel and characterise the novel aspects of the process.

All production steps must be documented and comprehensively described in a novel food application. The applicant must list all input materials, their functional role and the production process steps at which they are used. Input materials must be identified with a chemical identifier, certificates of analysis and regulatory status in the EU. For fermentation, the applicant must specify the quality of input/raw materials and fermentation aids. Production yield should be calculated.[143]

The information includes measures of production control and quality and safety assurance (e.g. HACCP principles in line with Regulation (EC) No 852/2004, GMP, ISO) and standardisation criteria (chemical markers for the novel food). Key steps and parameters of the production process, including information on potential impurities, by-products, impurities, contaminants, operational limits should be described in detail.[144]

A flow chart of the production process can be included in the application dossier as well, including quality and safety control checks.

Information should also be provided on the handling of the sources, for example, the propagation growth and harvesting conditions for fungi (e.g. wild or cultivated, time of harvest, type of cultivation and cultivation practices which also include information on the use of pesticides, hormones, antimicrobials and antiparasitic agents). Biological agents that can infect the fungi or its tissue cultures must be considered in the assessment.[145]

Post-harvesting handling, e.g. transport, drying techniques and storage conditions (duration, light, moisture and temperature) of unprocessed foods and the raw material should be specified and information on other starting substances or materials should be provided.

Culture conditions and growth medium should be specified. Also, it should be explained in detail how the raw material (a macroscopic fungi) is converted to an ingredient, or a preparation intended for a food product. Examples of processes may include heat treatment, extraction, distillation, squeezing, fractionation, purification, concentration, fermentation, chemical synthesis, enzyme-catalysis, or isolation from a natural source or others. Information on substances used in the manufacturing process, e.g. identity of the extraction solvents, ratio of extraction solvent to the material, reagents, residues, etc. should be provided.[146]

The techniques used to remove or inactivate microbial cells during downstream processing should be described in detail, including operational conditions. On the

---

143 EFSA Panel on Nutrition, Novel Foods and Food Allergens (2024a).

144 EFSA Panel on Nutrition, Novel Foods and Food Allergens (2024a).

145 EFSA Panel on Nutrition, Novel Foods and Food Allergens (2024a).

146 EFSA Panel on Nutrition, Novel Foods and Food Allergens (2024a).

other hand, where a novel food consists of viable cells, operational conditions ensuring the viability should be provided. For example, in the case of *Lentinus edodes* extract as a novel food ingredient, the EFSA Panel concluded that the employed technology based on submerged cultivation of mycelium in sterilised liquid medium enable the reproducible and standardised production of the novel food. The culture conditions were considered unlikely to lead to the production of secondary metabolites. From the safety perspective, it is also interesting to remark that the Panel noted the presence of soy peptides in the medium culture.[147]

For cell culture or tissue culture derived from macroscopic fungi, the applicant must provide the information on the type of cells used as source, information on the source, purification steps, cell isolation, selection, subculture, absence of pathogens and microbial contaminants must be provided.[148]

The employed technology may further affect novel food labelling, provided that the authorisation is successful. For example, UV-treated mushrooms *Agaricus bisporus*, the designation on the label of the novel food as such or of the foodstuffs containing it must be accompanied by indication that a controlled light treatment was used to increase vitamin D levels or UV treatment was used to increase vitamin D2 levels.

3.14 The Influence of Growth Media and Cultivation Conditions

The genetic regulation of biosynthetic genes, i.e., genes governing mycotoxin formation, is often complex and depends not only on intrinsic factors but also on various extrinsic factors, such as the composition of the growth medium and other cultivation conditions like temperature, oxygen availability or humidity. Currently, it is impossible to predict which substrates or under which growth parameters biosynthetic genes will be activated ("expressed") or remain inactive ("silent"). This fact seriously complicates the safety assessment of foods produced with the help of filamentous fungi. For instance, the mould *Penicillium roqueforti*, a filamentous fungus used for the maturation of blue cheese such as Roquefort or Gorgonzola, is safe when used in the cheese-making process.[149] Both the substrate as well as controlled fermentation conditions ensure that mycotoxins are produced in trace amounts only and partially react with the cheese matrix, further reducing the concentration of these compounds.[150] However, under certain conditions, such as when the mould grows on silage or other substrates, it can produce various mycotoxins like roquefortines, mycophenolic acid, or PR toxins in sufficient amounts to pose

147 EFSA (2010b).

148 EFSA Panel on Nutrition, Novel Foods and Food Allergens (2024a).

149 Coton et al. (2020); Scott (1981).

150 Coton et al. (2020); Scott (1981).

health risks on the consumer.[151] Therefore, it is reasonable that fermentation on novel growth media or under significantly altered physical process parameters trigger the novel food status and require individual safety assessment. This is especially important for fungi with a complex biosynthetic potential. A different mechanism that also potentially compromises the safety of MMPs in different growth media is the capacity of filamentous fungi to absorb and enrich compounds from the environment such as toxic heavy metals.

3.15 Proposed Uses

As part of the premarket authorisation dossier, applicants should further specify the purpose of how the food is intended to be used (e.g. ingredient, whole food).[152]

Proposed uses need to be expressed according to the EFSA Food Additive Intake Model (FAIM) tool or the Dietary Exposure (DietEx) tool. Broad categories should be used. Proposed uses must be given for every form of novel food with maximum concentration levels (see Table 3.5).

A novel food has a certain target population (adults, children, the general population, etc.) that should be cross-referenced with relevant safety data. Also, where certain population subgroups might not be able to avoid consumption of the novel food, that data should also cover those groups.[153]

The applicant should also propose the maximum amount of consumption of the novel food, together with the proposed average and maximum daily intakes for different age/gender/target population groups as appropriate (per kg body weight and in absolute amounts). The applicant should state the form and/or concentration of the product (fluid, extract, etc.) and if the ingredient is present in a food supplement, the applicant should also state possible indication of the quantity/amount.[154]

Chronic intake estimates should be provided too[155] and should document sources of data used, scientific principles and methods applied, in particular regarding the model. Mean and the 95th percentile anticipated intake level should be provided, also combining intake scenarios resulting from different uses of the novel food. Acute intake levels should be provided if toxicological, human data, data on chemical composition or literature review raise acute effect concerns. The applicant may also consider combined exposure to constituents in a novel food with additional sources of the constituents, such as in case of vitamin D2-enriched mushroom powder. A combined intake would thus consider:

[151] Scott (1981).

[152] EFSA Panel on Nutrition, Novel Foods and Food Allergens (2024a).

[153] EFSA Panel on Nutrition, Novel Foods and Food Allergens (2024a).

[154] EFSA Panel on Nutrition, Novel Foods and Food Allergens (2024a).

[155] EFSA Scientific Committee (2012). The methodological aspects of the intake assessment should consider the EFSA guidance on default values and rounding.

Table 3.5 Examples of specifications as retrieved from Commission Implementing Regulation (EU) 2017/2470

Novel food	Specified food category	Maximum levels (per 100 g)	Labelling
Arachidonic acid-rich oil from the fungus *Mortierella alpina*	Infant formula and follow-on formula as defined in Regulation (EU) No 609/2013 Foods for special medical purposes for infants as defined in Regulation (EU) No 609/2013	In accordance with Regulation (EU) No 609/2013	The designation of the novel food on the labelling of the foodstuffs containing it shall be 'Oil from *Mortierella alpina*' or '*Mortierella alpina* oil'
Pea and rice protein fermented by *Lentinula edodes* (Shiitake mushroom) mycelia	Bakery wares, breads Breakfast cereals Fruit and veggie drinks Ready to mix powders Chocolate confectionary Dairy analogues Fermented milk products Pasta products Meat preparations Soups Salads Meat analogues Milk-based drinks Single meal replacement	5 g 33 g 20 g 93 g 7 g 11 g 5 g 15 g 14 g 3 g 26 g 1 g 1 g	The designation of the novel food on the labelling of the foodstuffs containing it shall be 'Pea and rice protein fermented by Shiitake mushroom mycelia

- High daily intakes of the novel food and maximum use levels
- Mean and high daily intakes from natural sources derived from literature (background diet)
- Daily intake from other uses, including non-dietary sources (e.g. consumer products and pharmaceuticals, food fortification and supplements).[156]

Any information on whether the novel food is intended to replace another food should be appended. To estimate the anticipated intake level of such food, the applicant should use the consumption data of food already consumed in the EU, including natural occurrences of the novel food.[157]

The applicant should also explain whether the novel food is intended to replace another food, and whether it is reasonable to expect that the novel food would succeed in doing so. This is particularly relevant for mycelium-fermented proteins that are considered a type of alternative proteins (substitutes for meat, fish, dairy, etc.)

[156] EFSA Panel on Nutrition, Novel Foods and Food Allergens (2024a).

[157] EFSA Panel on Nutrition, Novel Foods and Food Allergens (2024a).

and where the comparison of the protein content could be of relevance. Protein content should be quantified using the 6.25 nitrogen-to-protein conversion factor and as the sum of the anhydrous amino acids. The complete amino acid profile should be submitted. If novel foods are enriched with specific proteins, the applicant must characterise the individual protein.[158]

Conservative scenarios should be used.[159]

3.16 Composition and Stability of Novel Foods with Agri-Food Side Streams as Substrates

Fermentation with mycelium as a fermentation agent needs carbon sources for biomass formation. Biomass is developed by degrading sugars (substrate, such as high-refined glucose syrup) under optimal conditions. Mycelium can be used in the fermentation of grains (corn, wheat), legumes, apples, wood, or other substrates, sometimes in combination. Currently, agricultural by-products are considered an alternative source of carbon. Fermentation may proceed as solid-state or submerged, although in submerged fermentation, the conditions require greater precision than in solid state fermentation.[160] Also, several companies strive to optimise the production process in a way that the resulting biomass is minimally processed. The crucial consideration is the conversation ratio between the carbon sources and the final protein product in terms of its nutritional content and protein quality.[161]

The choice of the substrate may fundamentally alter the characteristics of the final product. From the safety perspectives, allergens, mycotoxins and heavy metals need to be primarily considered.[162] Many laboratory experiments and proofs of concept have been dedicated to that matter. Metabolism of fungi may produce different composition of the final product. Therefore, qualitative and quantitative information on chemical, physicochemical, microbiological and nutritional attributes of the novel food need to be considered with every change of a substrate.

Compositional analysis should reflect variability in production processes and their parameters. Another important consideration is the product standardisation. The biochemical quality of side streams, which might be used as substrates, may differ substantially making it challenging to arrive at a stable product. It is necessary to consider the complexities of storage, sterilisation and the avoidance of contamination. Also, side streams have limited shelf life. For these purposes, one must bear in mind the requirements for the novel food application concerning the product stability that need to be submitted as a part of novel food application:

---

[158] EFSA Scientific Committee (2012).

[159] EFSA Scientific Committee (2012).

[160] Kirsch et al. (2016).

[161] Zwinkels et al. (2023).

[162] Hadi and Brightwell (2021).

- Stability test on preferably at least five, representative, independently produced batches of the novel food, produced at an industrial production scale or at one representative of it.
  - Test duration must cover at least the end of the novel food shelf-life with possible intermediate intervals of testing
  - Stability tests need to be performed to identify hazards which arise during storage and transports. This entails any monitoring constituents and parameters susceptible to changes during storage and having direct effect on safety, and to effects of packaging, temperature and the environment. Often, proposed conditions of use are not considered and addressed properly.[163]
- Characterisation of the nature of degradation products
- Information on normal storage conditions of the novel food as well as information on the storage conditions (temperature, light exposure, oxygen, moisture) under which the stability test was performed
- Physicochemical, biochemical and microbiological stability of the novel food under normal conditions of storage
- If the novel food is used as an ingredient added to other foods the manufacture of which requires further processing, the impact of this processing on the novel food must be examined. Characterisation of the stability of the novel food in real food matrices or in relevant model systems (e.g. mimicking the food matrix and the respective processing conditions such as the effect of processing temperature, pH)
- Information on ingredients added to the novel food to improve stability, if relevant.[164]

For the inclusion of agricultural by-products in production processes, one must also consider general and specific hygiene requirements per Article 4 of Regulation (EC) No 852/2004[165] on the hygiene of foodstuffs as well as applicable national laws and regulations. Agricultural or food by-products in general may be considered a contamination hazard and be subject to good hygiene practice. This is usually done in the process of sterilisation or pasteurisation. Food business operators carrying out primary production and those associated operations (e.g. storage) must comply with general hygiene provisions laid down in part A of Annex I and any specific requirements provided for in Regulation (EC) No 853/2004.[166] Food business operators must, as appropriate, adopt specific hygiene measures, Hazard Analysis and Critical Control Points.[167] This may also impact the use of agricultural by-products in compliance with EU law.

[163] Heinonen and Mangelsdorf (2017).

[164] EFSA Panel on Nutrition, Novel Foods and Food Allergens (2024a).

[165] Regulation (EC) No 852/2004.

[166] Regulation (EC) No 853/2004.

[167] van der Fels-Klerx et al. (2024).

Commission Regulation (EC) No 2073/2005 of 15 November 2005 on microbiological criteria for foodstuffs also needs to be observed.

Agricultural waste is not an agricultural by-product. Per Article 5 of Directive 2008/98/EC[168] on waste, Member States must take appropriate measures to ensure that a substance or object resulting from a production process the primary aim of which is not the production of that substance or object is considered not to be waste, but to be a by-product if the following conditions are met:

(1) further use of the substance or object is certain: This means that it is guaranteed that the material will be used, e.g. to meet the needs of economic operators other than the economic operator which produced it.[169]
(2) the substance or object can be used directly without any further processing other than normal industrial practice: Those treatment techniques that address typical waste-related characteristics of the production residue, such as its contamination with components which are hazardous or not useful, would prevent classification as non-waste. On the other hand, if a production residue is treated with normal industrial practice, e.g. modification of size or shape by mechanical treatment, it can be regarded as a by-product.
(3) the substance or object is produced as an integral part of a production process; and
(4) further use is lawful, i.e. the substance or object fulfils all relevant product, environmental and health protection requirements for the specific use and will not lead to overall adverse environmental or human health impacts. It is important to note that where an agricultural by-product is used as fermentation or cultivation substrate, and the resulting product requires a novel food authorisation, the by-product does not fulfil all preconditions in the absence of such an authorisation and would be likely considered waste. A producer of mushroom or mycelium products should verify whether the substrate supplier complies with the regulatory requirements for by-products. These tests are cumulative. A decision on whether a particular substance or object is a by-product must be made by the producer of the substance or object, together with the competent national authorities, based on the applicable national legislation transposing the Directive 2008/98/EC.

[168] Directive 2008/98/EC of the European Parliament and of the Council of 19 November 2008 on waste and repealing certain Directives, OJ L 312, 22.11.2008, pp. 3–30.

[169] European Commission (2008).

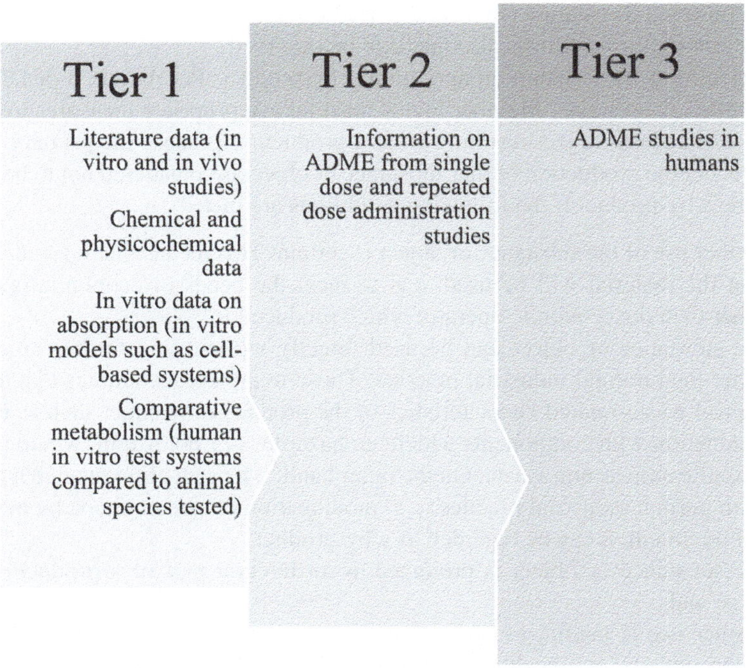

Fig. 3.3 ADME testing. Source: Adapted from EFSA 2024 Novel Food Guidance [EFSA Panel on Nutrition, Novel Foods and Food Allergens (2024a)]

3.17 Absorption, Distribution, Metabolism and Excretion (ADME)

ADME is relevant for nutritional and toxicological assessment. It informs about how the novel food is absorbed, how and to what extent it is metabolised and eliminated and so not accumulated. Potential matrix effects should be considered. ADME might not be necessary in case of novel foods composed of substances commonly found in the body or in the diet. ADME is to be conducted following a tiered approach (see Fig. 3.3)[170]

- If there is an indication that the novel food or its components are not absorbed in the small intestine, *in vitro* studies mimicking the human gut and its microbiota dynamics should be conducted to identify and quantify metabolites of concern.
- If Tier 1 data is not sufficient or indicates that the novel food or its constituents are absorbed or systematically available or accumulated in the body or metabolites of concern are formed, the applicant must resort to Tier 2 or 3. Tier 2 would

[170] EFSA Panel on Nutrition, Novel Foods and Food Allergens (2024a).

be also required where a prolonged half-life or enzyme induction is observed or expected.

- If there are substantial differences in comparative *in vitro* metabolism studies between different species or test species and humans or evidence for bioaccumulation in the tested species, the applicant must perform ADME studies in humans.
- Bioavailability (non-mandatory) must be tested only when the novel food is also a new nutrient source. If the novel food is a source of new micronutrient, bioavailability needs to be quantified. If it is, bioavailability needs to be assessed and demonstrated but not quantified. Where there is a potential concern about a protein in the novel food, protein digestibility studies should be performed as part of the weight of evidence approach for the nutritional, toxicological and allergic assessments.[171]

3.18 Nutritional Information

Nutritional information provides evidence on whether a novel food is nutritionally disadvantageous for consumers under the proposed conditions of use. The assessment concerns nutrient and non-nutrient components with an established role in the diet for which dietary reference values have been established by the EFSA.[172] Three scenarios could be considered:

1. Comparison of anticipated intakes of nutrients from the novel food vs health-based guidance values (tolerable upper intake levels) and/or intakes in EU populations
2. Comparison of composition data regarding antinutrients vs comparable sources of antinutrients, considering anticipated intake levels
3. In case of novel foods intended to replace other foods in the diet, comparison of the nutrient composition of the novel food with a reference food, considering anticipated intake levels. This includes novel food production process that is susceptible of affecting the nutrient composition.[173]

The applicant must establish that the novel food cannot cause excess intake of nutrients, i.e. exceeding upper limits or other health-based guidance values or cannot adversely affect the food nutritional status of consumers by increasing the risk of inadequate nutrient intakes. Other scenarios lead to the conclusion that the consumption of the novel food is nutritionally disadvantageous. In that case, the applicant may consider modifying the novel food's composition and/or conditions of use.

The following factors may be considered:

[171] EFSA Panel on Nutrition, Novel Foods and Food Allergens (2024a).

[172] Here, the EFSA developed a special tool - the DRV Finder that gives quick and easy access to EFSA's Dietary reference values (DRVs) for nutrients.

[173] EFSA Panel on Nutrition, Novel Foods and Food Allergens (2024a).

- Target groups, maximum specification limits and proposed use levels, including combined nutrient intakes.
- Bioavailability of nutrients, considering influences of the production process, handling storage and further processing prior to consumption. Data from in vitro, in silico, in vivo human or animal studies may need to be considered regarding the interaction of the novel food with the diet and nutrients.
- Content and effect of antinutritional factors (tannins, lectins, trypsin inhibiters, etc.) in the novel food as compared to antinutrients in comparable food.
- Focus on nutrients that represent a significant source in diets (indicatively, 15 g per 100 g or 100 mL).[174]

As regards mycelium derived proteins, the protein quality must be assessed if the highest mean consumption of the novel food under the proposed conditions of use could substantially contribute to the average requirement for protein for one or more population groups (by at least 15 percent). Protein quality depends on several factors, including the content of indispensable amino acids, the presence of antinutrients and processing. Protein quality is to be assessed using the Digestible Indispensable Amino Acid Score (DIAAS) method, which is regarded as the gold standard for protein quality assessment. This method involves *in vivo* digestibility studies in established animal models and humans. While *in vitro* models can approximate *in vivo* digestion, they do not fully capture the complexity of *in vivo* processes and have yet to be validated for regulatory purposes.[175]

3.19 Toxicology

Toxicological studies identify and characterise potential hazards and novel foods and help establishing safe intake levels. The assessment of safety should be based on all available information that supports the safety of the novel food under the proposed conditions of use. The information helps to establish the need for toxicity studies and the corresponding testing strategy. The assessment considers the compositional data and ADME of the novel food and is conducted against the backdrop of a comprehensive literature review (toxicological information of the novel food, its constituents or metabolites, human studies and case reports, information and safety assessment from non-food uses). In light of this information, the outstanding data gaps on hazard identification and characterisation must be filled by an appropriate testing strategy with a testing material derived from the production process with the submitted compositional data and specifications. The assessment must also consider sources of uncertainties.[176]

The applicant may consider:

[174] EFSA Panel on Nutrition, Novel Foods and Food Allergens (2024a).

[175] FAO (2022).

[176] EFSA Panel on Nutrition, Novel Foods and Food Allergens (2024a).

- Read-across in case of single mixtures, focusing on toxicological data on structurally related substances and their metabolic profiles
- Threshold of toxicological concern (TTC) approach as regards substances at low exposure levels such as impurities, metabolites and degradation products
- QPS approach to safety assessment.[177]

Rationale for the toxicity testing strategy applied should be provided. A testing strategy depends on the type of test material (single, simple mixture versus complex mixture versus whole foods that should be tested like complex mixtures).[178]

3.19.1 Genotoxicity

Assessing genotoxicity is considered basic chemical risk assessment in food. In case of fungi used as a novel food, either as active agents or biomasses, the applicant may perform genotoxicity depending on the taxonomic classification and hazard identification of the microorganism. This is tested using the supernatant (from at least three independently produced batches of the novel food pooled together) and the cell lysate (from at least one batch of the novel food). Proof of efficient lysis of the cells/spores must be submitted. The assessment is based on tiered approach (see also Fig. 3.4):

1. **Tier I**: basic battery of in vitro tests (bacterial reverse mutation assay and in vitro micronucleus assay). In case of novel foods made of proteins, special testing may be needed, such as mammalian gene mutation assays or bacterial reverse "treat and plate" methodology
   a. In case of positive in vitro micronucleus test, the applicant must perform test on whether the novel food exhibits aneugenic potential, by conducting a kinetochore staining or fluorescence in situ hybridisation
2. **Tier II**: positive or ambiguous findings suggesting potential genotoxicity in Tier I triggers the need to conduct *in vivo* genotoxicity tests of substances present in food and provide evidence of target tissue exposure.[179]

---

[177] EFSA Panel on Nutrition, Novel Foods and Food Allergens (2024a).

[178] According to the Scientific Panel on Food Additives and Nutrient Sources, complex mixtures are "conventional metabolism and toxicokinetic studies may not be feasible for all components in the mixture but should be provided for toxicologically relevant constituents. Toxicologically relevant constituents are generally considered to be the major components and those other components with known or demonstrable biological or toxicological activity, and should be determined on a case-by-case basis with a scientific justification and the rationale for their selection provided".

[179] EFSA Panel on Nutrition, Novel Foods and Food Allergens (2024a).

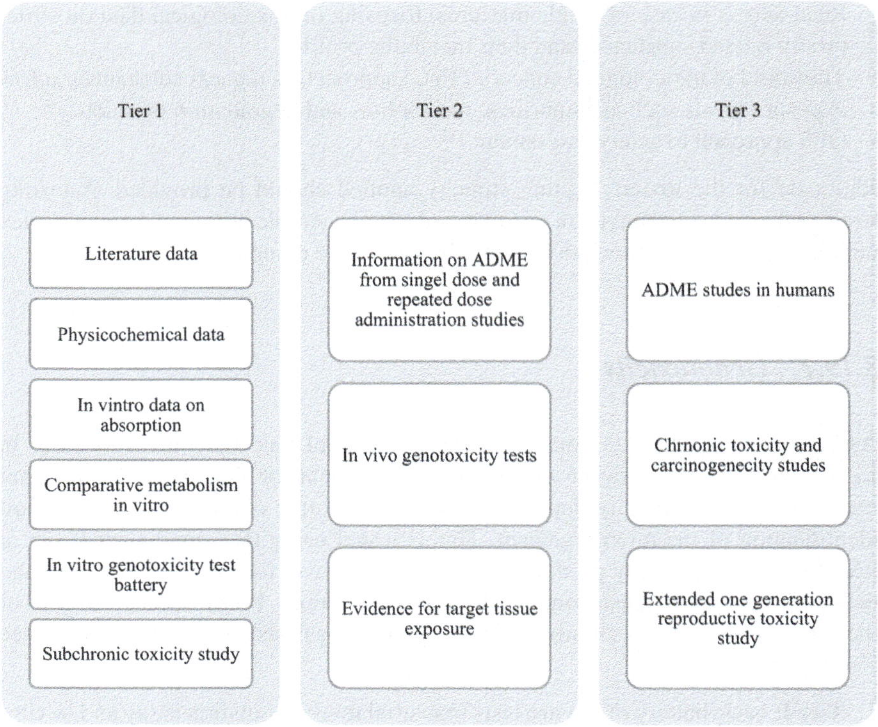

Fig. 3.4 Tiered approach to toxicity assessments [EFSA Panel on Nutrition, Novel Foods and Food Allergens (2024a)]

3.19.2 Repeated-Dose Toxicological Studies

1. **Tier I:**
    a. Subacute toxicity: may be performed providing the basis for the selection of appropriate doses to be used in subchronic setting
    b. Subchronic toxicity: i.e. 90-day study for the novel food of unknown toxicity where no health-based guidance values are available for the components. The study should allow for the determination of the lower bound of the benchmark dose or a no-observed-adverse-effect-level. Uncertainty factor of 200 is applied to the reference point identified in the study. A rationale for the selected administration route in consideration of the proposed conditions of use should be provided. The applicant should submit an overview table of study reports and summary table of statistically significant and dose-dependent observations in toxicity studies, summarising the statistically significant differences between controls and the novel food
        i. Relevant historical control data for the species/strain should be included.

   ii. Satellite groups for possible toxicokinetic determinations should be considered.
   iii. Optional endpoints should be considered in case of endocrine, reproductive or cardiovascular concerns such as oestrous cycling for at least 15 days prior to necropsy, sperm quality analyses, histopathological analyses or reproductive organs and hormonal measurements at necropsy.

2. **Tier II**

   a. Mechanistic and investigative studies to assess specific toxicity
   b. Reproductive, endocrine and developmental toxicity studies are performed based on kinetic and toxicity data, including read-across data from structurally related compounds and literature data. Additional studies include OECD TG 414, 416, 421, 422, 426, 440, 441, 455, 456 and 493

      i. The need for other studies, such as on neurotoxicity, cardiovascular effects, immunotoxicity, hypersensitivity and food intolerance may be triggered by positive findings in the literature of Tier I or II.

3. **Tier III**:

   a. Extended one-generation reproductive toxicity study covers reproductive/developmental toxicity, neurotoxicity and development immunotoxicity and is triggered by findings in the literature, *in vitro*, *in vivo* and human studies.
   b. Chronic toxicity and carcinogenicity are normally not required but may be required due to accumulation of the novel food or hyperplasia observed in subchronic toxicity studies.[180]

- Human data may be required to further assess potential adverse effects of the novel food or for effects that cannot be investigated in animals (e.g. psychological outcomes, mental health) and for novel foods that are new sources of micronutrients:

  – Available human intervention, epidemiological and observational studies relevant to the safety assessment should be organised according to a hierarchy of study designs and research questions, reflecting the relative strength of evidence which may be obtained from different types of studies
  – Efficacy studies with the novel food, although not directly related to safety, may provide some information deemed safety relevant

  Physical examination, blood chemistry, haematology, urine analysis, blood pressure, organ function tests, monitoring of adverse reactions.[181]

Supporting evidence on the history of safe use can be also used in the safety assessment. For example, if it is relevant for the safety assessment, a comprehensive literature review of studies with specific and safety-related components of the novel food and for studies with similar foods from the same or other closely related

---

[180] EFSA Panel on Nutrition, Novel Foods and Food Allergens (2024a).

[181] EFSA Panel on Nutrition, Novel Foods and Food Allergens (2024a).

sources. All available information on previous human consumption of a fungus and its source as well as other information, such as kinetic data, anticipated use, non-food use is considered in order to decide which toxicity studies are necessary to be conducted to prove the novel food's safety.

As such, subchronic toxicity studies would be required for glucosamine from genetically modified *A. niger* but not for milk fermented with *Bacteroides xylanisolvens* with well characterised source with a significant history of food use, comprehensive compositional data, no concerns from production process, and knowledge on the main components.[182]

Toxicological studies must be conducted in facilities which comply with the requirements of Directive 2004/10/EC[183] or, if they are carried out outside the territory of the Union, they should follow the OECD Principles of Good Laboratory Practice. The applicant must provide evidence of compliance with those requirements and justify any deviation from the standard protocols.[184]

A certificate of analysis of the test material used in these studies must be provided.[185] If any of toxicological studies used a test material different from the novel food, an explanation must be provided that would explain why the test material used is appropriate for the safety assessment of the novel food. This is because toxicological studies should be in principle carried out with the novel food as intended to be marketed, i.e. the test material should be manufactured according to production process described and meet the compositional characteristics and the specifications provided.[186]

For each biological or toxicological study, the applicant must clarify whether the test material conforms to the proposed or existing specification. Where the test material differs from that specification, the applicant must demonstrate the relevance of those data to the novel food under consideration.[187]

(Quantitative) structure activity relationship is also relevant ((Q)SAR).[188] It is suggested that findings from QSAR analyses can be further used as input

---

[182] Heinonen and Mangelsdorf (2017).

[183] Directive 2004/10/EC.

[184] Heinonen and Mangelsdorf (2017).

[185] Heinonen and Mangelsdorf (2017).

[186] Heinonen and Mangelsdorf (2017).

[187] Specifications include:

- the limits and information on the exact method for each of the selected parameters (i.e., as a minimum, the contents and/or limits for the parameters on the identity of the product, the minimal purity, limits acceptable for impurities and degradation products, in particular those of toxicological or nutritional relevance. In the absence of legal requirements in the EU, it also includes maximum levels of contaminants)
- the limits and information on nutritional or biologically active components or, when these are not known, on selected chemical markers-the limits and information on concentrations of the major group of constituents (e.g. amino acids, proteins, etc.)

[188] Heinonen and Mangelsdorf (2017).

parameters for physiologically based toxicokinetic modelling (PBTK).[189] Toxicological data on structurally related substances (read-across) should be considered in case of insufficient other data. The EFSA Compendium of Botanicals contains a database of plants reported to contain naturally occurring substances of possible concern. It is used to identify adverse health effect with specific species and may be used for read-across data. The compendium currently does not include algae, cyanobacteria and fungi; they will be considered for possible inclusion in the future.[190]

3.20 Special Toxicological Considerations Concerning Mycelium

Mycelium and fruiting bodies produce different toxins. For instance, fruiting bodies of the hallucinogenic "magic" mushrooms (genus *Psilocybe*) contain the high amounts of psychoactive compounds, while mycelium raised in liquid culture, however, is a poor producer of the compounds.[191] Generally, it is hypothesized that bioactive compound formation often serves as defence mechanism, specifically protecting the fruiting body, a structure particularly vulnerable to grazing animals. This may be the reason why fruiting bodies are often more productive in terms of toxin production than the vegetative mycelium itself. However, this is not generally true. An example for the opposite situation is *Armillaria mellea* (honey mushroom), whose fruit body is regarded as edible after thorough cooking. This species is a parasitic fungus that causes significant damage in orchards and is also able to grow on decaying wood. Its vegetative mycelium produces high amounts of melleolides, a complex group of cytotoxic compounds. The fruiting body, however, contains only small amounts of these compounds.[192] When discussing the differences between mycelia and fruit bodies, it should be emphasized that there is no standard mycelium or a standard fruit body of a species. Since each structure might behave differently under varying cultivation conditions, statements about the difference between mycelia and fruit bodies cannot easily be generalized.

---

[189] de Boer and Bast (2018).

[190] EFSA (2021a).

[191] Waldbillig et al. (2023).

[192] Dörfer et al. (2019).

3.21 Allergenicity

The potential to cause allergic reactions is one of the main hazards associated with proteins produced by mycelium. Allergenic potential of the novel food must be tested, considering its composition, particularly its protein, its source (including taxonomic relationships), the production process and available experimental and human data, including information on cross-reactivity.

If the novel food is produced exclusively from/with materials not containing proteins, no data are required as regards allergenicity.[193]

If the novel food is produced from/with materials containing allergic foods listed in Annex II Regulation (EU) No 1169/2011, the novel food is subject to mandatory labelling requirement.

If the novel food is produced from/with materials known to trigger allergic reactions in susceptible individuals, but which are not listed in Annex II, the allergenic potential of the novel food may be known based on the information from literature or unknown. In the former case, the applicant must generate data on detection and identification of the known clinically relevant allergenic proteins. The applicant is further required to submit the following information is available in the literature:

- Prevalence of the food allergy related to the novel food source
- Type and severity of symptoms triggered by the novel food source
- Potency of the allergic novel food source, e.g. from the clinical history
- Known clinically relevant allergenic proteins of the novel food source

In the latter case, the applicant must submit the following information:

- A comprehensive literature review in order to retrieve available information on allergenic potential of the novel food, its source, used raw materials, other products from the same source and closely related species to the source: case reports of allergic reactions, allergenicity studies (*in vitro*, in animals, in humans) of the novel food and/or its source, reactivity, elicitation dose, sensitisation and clinical effects
- Protein digestibility and/or stability
- Information on cross-reactive allergenicity. This is to confirm whether individuals allergic to the known food allergen which shares high amino acid sequence homology with the novel food protein could also have allergic reaction to the novel food. First, comprehensive literature review on the allergenicity of the novel food source organism and its phylogenetic relationship with known allergenic sources should be performed.[194]

---

[193] EFSA Panel on Nutrition, Novel Foods and Food Allergens (2024a).

[194] EFSA Panel on Nutrition, Novel Foods and Food Allergens (2024a).

If the literature review renders a positive result, the applicant must perform bioinformatic search for cross-allergenicity. By default, amino acid sequence homology comparison should be performed.[195]

If results for cross-allergenicity are positive, a follow up analysis with human serum sourced from allergic individuals with a confirmed allergy to the known allergen with which the novel food may cross-react, should be performed. To assess cross-reactivity, immunoassay methods, such as ELISA, are adequate.[196]

If cross-allergenicity was further confirmed, it necessitates further investigation in humans. Then, prick test and double bling placebo-controlled food challenge studies are required.

Other appropriate methods to support the allergenicity investigation of novel foods:

- Protein content in the novel food (including limit of detection and quantification) and accurate description of the methods used
- Molecular weight of the potentially allergenic protein, heat stability, sensitivity to pH, digestibility by gastrointestinal proteases
- Degree of sequence homology with known allergens.[197]

Allergenicity testing is particularly important for proteins produced via mycelium fermentation. The default assumption for novel foods containing proteins is that they have allergenic potential. This is due to a risk of *de novo* sensitisation or cross reactivity.[198] However, the applicant wishing to prove that the novel food is unlikely to trigger an allergic reaction should follow the EFSA guidance on the preparation and presentation of applications pursuant to Article 21 Paragraph 2 of Regulation (EU) No 1169/2011, as amended.[199] Alie de Boer suggests detecting a potential allergen in four phases:

1. "Collect information about the history of exposure to the protein, while taking into account environmental and geographical factors
2. Observe the taxonomy of the novel food proteins and the relationship between these proteins and known allergens to indicate potential allergens
3. Identify the novel food proteins and compare them with proteins in specific databases
4. Consider the way that consumers will use the product because of potential matrix changes due to processing and preparation of the novel food and its influence on the putative allergenic potential"[200]

---

[195] Using FASTA local alignment algorithm or the BLAST algorithm, with complementary searches based on 3D protein structure or machine learning strategies based on IgE etipole matching.

[196] EFSA Panel on Nutrition, Novel Foods and Food Allergens (2024a).

[197] EFSA Panel on Nutrition, Novel Foods and Food Allergens (2024a).

[198] de Boer and Bast (2018).

[199] EFSA Panel on Dietetic Products, Nutrition and Allergies (2013).

[200] de Boer and Bast (2018).

3.22 Potentially Harmful Traits of Filamentous Fungi

The "Potentially harmful traits of some strains of the taxonomic unit", formulated by the EFSA, are typically bioactive metabolites, nucleic acids, allergens or toxic compounds taken up from the environment that exert undesired effects on human health. Microorganisms, especially filamentous fungi often harbour a huge genetic potential for the production of bioactive metabolites. These compounds, ranging from beneficial molecules like vitamins or antioxidants to poisonous substances, may exert various biological activities, dependent on their dose and mode of action. Toxic metabolites produced by fungi, generally known as "mycotoxins" are amongst the most significant chronic dietary risks that need to be controlled to safeguard public health.[201] The diseases and symptoms linked to mycotoxins range from gastrointestinal disturbances to liver and kidney damage, various cancers, neurological disorders or the suppression of the immune system, amongst others. While some effects of mycotoxins, e.g. of typical mushroom poisons, comprise acute toxicity, i.e. severe health effects shortly after a single exposure to the compound, they are often subchronic (mid-term effects, after repeated exposure over a period of weeks) or chronic (long-term effects, typically after prolonged and repeated exposure). While acute toxicity is unlikely to go unrecognized in food production for long, the identification of chronic toxicity requires careful monitoring over an extended period. The most prominent example of mycotoxins are the aflatoxins, a family of closely related compounds produced by the filamentous molds *Aspergillus parasiticus and A. flavus,* from which their name is derived. Aflatoxins are a common contaminant of nuts, cereals and various other foods. In high doses, aflatoxins cause acute toxicity involving liver damage and subsequent conditions such as hepatic encephalopathy, a brain disorder.[202] However, even in relatively low concentrations, aflatoxins are highly cancerogenic to humans causing especially liver cancer. Because of their extremely high cancerogenic potential, aflatoxins are amongst the most dangerous food contaminants. Researchers have gathered a considerable body of knowledge about mycotoxins, their health effects but also their biosynthesis by filamentous mold.[203] Notably, some fungal species can produce several mycotoxins. At the same time, certain mycotoxins are produced by various species. Each bioactive metabolite is formed through a dedicated metabolic pathway. Their corresponding biosynthetic genes are often organized in biosynthetic gene clusters.[204] The presence of well-known biosynthetic genes, such as the aflatoxin gene cluster, indicates an organism's potential to produce the corresponding compounds under certain conditions. Conversely, the absence of a specific mycotoxin gene cluster in an organism's genome may rule out the risk of its production, especially when the biosynthetic pathway is complex. For instance, shiitake mushrooms cannot produce

[201] Richard (2007).

[202] Kensler et al. (2011).

[203] Patriarca and Fernández Pinto (2017); Richard (2017).

[204] Li et al. (2016).

aflatoxins because their genome lacks the necessary biosynthetic genes. However, caution is needed for less complex molecules, as alternative routes for their formation might exist.

In addition to bioactive compounds like mycotoxins, further health concerns were discussed in the context of fungi-derived foods. For instance, the mold *Fusarium venenatum* A3/5 (formerly known as *F. graminearum* A3/5) has been developed since the 1960s to produce fungal protein (mycoprotein) as an alternative to meat and other classical protein sources. While the protein was of high nutritional value, the product contained high amounts nucleic acids compared to products of animal or plant origin.[205] Nucleic acids such as RNA are metabolized to uriate (uric acid) in humans, a poorly soluble metabolite. Increased levels of uriate can cause gout due to precipitation of uriate crystals in joints and other tissues. This potential problem was overcome by removal of RNA during the production process.[206]

The occurrence of allergens in foods is another general health risk. Common food allergens occur in milk, eggs, fish, or nuts and many others, present in many types of foods. Labelling requirements help consumers with food allergies make safe choices and avoid potentially life-threatening reactions. Fungi may also contain allergens such as fungal proteins[207] or cell-wall polysaccharides. For instance, people suffering from allergies against molds have cross-reacted to products derived from other mold species - even distantly related ones.[208] Generally, such cases are relatively rare,[209] but allergic reactions to novel MMP are likely to occur and need to be rigorously investigated to confirm the allergenic potential. Products will require labelling when the allergenic potential has become evident.[210] However, the allergenic potential of novel foods is not restricted to fungal products but also affects plant-based products or proteins from unconventional sources such as insects and algae.[211]

Another safety concern is the heavy metal content of filamentous fungi. Edible mushrooms are known to accumulate heavy metals from the environment in their fruiting bodies.[212] In case of trace elements such as zinc (Zn), selenium (Se), this may have beneficial effects on consumer health, as long as the recommended daily allowance is not exceeded.[213] However, when toxic metals such as lead (Pb), cadmium (Cd) or arsenic (As) are accumulated, the product can no longer be regarded as safe.[214] The risk of uncontrolled heavy metal accumulation in cultivated fungi

---

[205] Wiebe (2002); Edelman et al. (1983).

[206] Wiebe (2002); Edelman et al. (1983).

[207] Hadi and Brightwell (2021).

[208] Katona (2002).

[209] Daba et al. (2021).

[210] Regulation (EU) No 1169/2011, Article 9.

[211] Hadi and Brightwell (2021).

[212] Dowlati et al. (2021).

[213] Rzymski et al. (2017).

[214] Dowlati et al. (2021).

might be especially high when poorly defined growth media such as agro-industrial waste products are used as growth substrates. However, it should be noted the risk of metal accumulation is not a particular problem of MMPs but also of plants[215] or sea food[216] for example.

3.23 Qualified Presumption of Safety

As discussed above, novel MMPs typically require an individual safety assessment to receive approval by the EFSA as novel foods, food additives or other category of regulated products. In addition to the choice of organism, changes in substrate or fermentation conditions are also likely to trigger the novel food status of MMPs. These regulations pose significant hurdles for innovators in the field, making it potentially challenging to bring their products to the market. A qualified presumption of safety (QPS) for filamentous fungi could offer potential relief in this situation. However, filamentous fungi are generally excluded from the qualified presumption of safety (QPS) since 2007 and continue to be excluded until today (2024).[217] This "general suspicion" against MMPs is criticized by some market participants and raised questions about why QPS status has not been granted at least for a subset of filamentous fungi, despite a growing body of knowledge about these organisms. In the following section, we will examine the scientific rationales behind the EFSA's careful approach to the approval of filamentous fungi. To justify the general exclusion of filamentous fungi from QPS assessment, the EFSA argues with "an ambiguous taxonomic position or the possession of potentially harmful traits by some strains of the taxonomic unit, therefore requiring a specific assessment for each strain for which an application is made."[218]

Since the EFSA recognises that a variety of fungal species are used in food and feed production, either directly (to produce fermented foods) or as additives, food enzymes and other components of food, specific administrative guidance has been provided for microorganisms, including yeasts species.[219] Some biological agents may be granted a QPS status that is established by the EFSA Panel on Biological Hazards. In 2007, the EFSA established a Qualified Presumption of Safety (QPS) list for microbial species used in food and feed production. This list has been supposed to work as a tool for prioritising within the risk assessment of those microorganisms which raise safety concerns.[220] It is constantly updated. The QPS is also not a criterion to decide on the novel food status. Fungi that do not have QPS status are

[215] Sevik et al. (2020).

[216] Stankovic et al. (2012).

[217] EFSA Panel on Biological Hazards (2024).

[218] EFSA Panel on Biological Hazards (2024).

[219] EFSA Panel on Additives and Products or Substances used in Animal Feed (2012).

[220] EFSA (2007).

Table 3.6 EFSA, Updated list of QPS status recommended biological agents in support of EFSA risk assessments

Yeasts (species)	Qualifications
Cyberlindnera jadinii	QPS applies for 'production purposes only' (the qualification 'for production purpose only' implies the absence of viable cells of the production organism in the final product and can also be applied for food and feed products based on microbial biomass).
Debaryomyces hansenii	
Hanseniaspora uvarum	
Kluyveromyces lactis *Kluyveromyces marxianus*	
Komagataella pastoris *Komagataella phaffli*	QPS applies for 'production purposes only'
Limtongozyma cylindracea	QPS applies for 'production purposes only'
Ogataea angusta	QPS applies for 'production purposes only'
Phaffia rhodozyma	
Saccharomyces bayanus *Saccharamyces cerevisiae* *Saccharamyces pastoranius*	
Schizosasscharomyces pombe	
Wickerhamomyces anomalus	QPS applies for production purposes only
Yarrowia lipolytica	QPS applies for production purposes only
Zygosaccharomyces rouxii	

not necessarily considered novel or vice versa. The QPS assessment is triggered internally by the EFSA upon receiving an application for novel food authorisation and cannot be triggered directly by an applicant. The list currently features the yeast species (members of the fungus kingdom) shown in Table 3.6.

In this regard it is important to note that filamentous fungi, bacteriophages, Streptomycetes, Oomycetes, *Enterococcus faecium*, *Escherichia coli* and recently also *Clostridium butyricum* are excluded from the QPS assessments based on an ambiguous taxonomic position or the possession of potentially harmful traits.[221] The EFSA panel stated: "While knowledge of fungal secondary metabolites has grown substantially, information on their toxic effects on humans and animals is evolving at a much slower rate. Therefore, it was decided that until further notice, filamentous fungi would be excluded from the QPS evaluations."[222] The EFSA stated that filamentous fungi are not included "based on an ambiguous taxonomic position or the possession of potentially harmful traits making it unlikely that any (taxonomic units…) would be granted the QPS status in the foreseeable future."[223] This is even

[221] EFSA Panel on Biological Hazards (2020).

[222] EFSA (2024b).

[223] EFSA Panel on Biological Hazards (2023).

though only between October 2019 and September 2022, the EFSA received 56 notifications concerning filamentous fungi to be included in the QPS list, however, none of the notifications was evaluated. The EFSA further stated that a case-by-case assessment at a strain level needs to be carried out by the relevant EFSA panel.[224] The distinction between a filamentous fungus and a yeast, and thus whether an organism would be eligible for QPS evaluation and status (yeast), or not (filamentous fungi) is sometimes not clear-cut.[225]

Yeasts are not expected to belong to any other fungal phylum than ascomycetes and basidiomycetes.[226] Of the three lineages within the Ascomycota, only two contain yeasts, whereas the third (*Pezizomycotina*) does not. The basic morphology of the *Pezizomycotina* is filamentous, and a reference is not made to any fungi within *Pezizomycotina* as "yeasts", or "yeast-like" either.[227] Typically, members of *Pezizomycotina* contain a high abundance of enzymes for secondary metabolism, which is generally in contrast to yeasts but similar to filamentous fungi.[228]

The decision whether a species should be considered to be a yeast or a filamentous fungus for QPS purposes is taken on a case-by-case basis, but applying the following general principles: A fungus may be subject to evaluation if it i) belongs to the phyla Ascomycota (excluding the *Pezizomycotina*) or Basidiomycota and ii) is treated as a yeast by taxonomic literature.[229] As supporting information, the taxonomy applied by internationally recognised microbial culture collections is considered.[230]

If a fungus (microorganism) is not assigned a QPS status, the decision tree in Fig. 3.5 was proposed on the safety assessment of such a fungus. Literature suggests also other methods for safety screening, such as the Threshold of Toxicological Concern which is calculated based on exposure data, chemical structure, metabolism and findings on toxicity. Exposure below the TTC level is considered to not present any safety concern and exceeding this level signals a need to further investigate safety issues of the specific compound. If this approach would be more suitable for MMP is a question open to debate.[231]

However, these are other methods that need to be explicitly recognised in the EFSA guidance documents to gain relevance.

During the QPS assessment of a microorganism, taxonomic identity, body of knowledge and potential safety concerns are assessed at a taxonomic unit level. The absence of antimycotic resistance should be proved if the yeasts are to be used as

[224] EFSA Panel on Biological Hazards (2023).

[225] EFSA Panel on Biological Hazards (2023).

[226] Kurtzman et al. (2011).

[227] Naranjo-Ortiz and Gabaldón (2019).

[228] EFSA Panel on Biological Hazards (2022).

[229] EFSA Panel on Biological Hazards (2022).

[230] See https://wi.knaw.nl/page/fungal_table for fungi strains and https://theyeasts.org/ for yeasts.

[231] de Boer and Bast (2018).

Strain characterisation and genome sequencing	Has the strain been characterised to genus and species level? **AND**	If NO – do it If YES – next step
	Has the strain been completely genome-sequenced?	
Screening for undesirable attributes and metabolites	Is the strain free from genetic elements encoding known virulance factors and/or toxins associated with pathogenicity? **AND**	If NO – additional safety studies required If YES – next step
	Is the strain free from of functional and transferable antibiotic resistence gene DNA?	
Gene modification	Has the strain been genetically modified?	If YES, additional safety studies required + authorisation IF NO – next step
Strain origin	Was the strain isolated from a source that has a history of safe consumption for which the species is a substantial and characterizing component (not simply an incidental isolate)? **AND**	If NO – additional safety studies required If YES – go on
	Has the species undergone comprehensive peer reviewed safety evaluation (e.g. QPS) or has it been affirmed safe by authoritiative group/qualified scientific experts? **AND**	
	Do findings published since completion of the peer review continue to support safety?	
Exposure levels	Will the intended use of the strain expand exposure of the species beyond the group that typically consume the species in foods in which it is typically found? **AND**	If NO – strain is considered safe If YES – additional safety studies are required
	Will the intended use of the stain expand overall intake of the species (e.g. increasing the number of foods beyond foods in which the speciesis typically found or using the strain as probiotic rather starter culture)?	
Additional safety issues	Does the strain induce undesirable physiological effects in appropriately designed safety studies (e.g. animal models or clinical trials)?	If NO – strain is considered safe If YES – strain is not appropriate for human consumption

Fig. 3.5 Adapted from: Pariza et al. Determining the safety of microbial cultures for consumption by humans and animals [Pariza et al. (2015)]

viable organisms in the food and feed chains.[232] The type and level of exposure, potential allergenicity to microbial residual components and hazards linked to the formulation or other aspects of the processing of those products are not assessed. The EFSA assesses these aspects separately in its safety assessments for market authorisation of novel foods.[233]

While it is well-established that various filamentous fungi harbour numerous genes to produce bioactive compounds, our knowledge about the toxicological potential of most of these fungi remains incomplete, thus preventing qualified presumption of safety for these species. One critical aspect that is still poorly understood is the fact that biosynthetic genes for potential toxins can be either 'active' or 'silent,' depending on the taxonomic unit and its environment.[234,235] For instance, the bioactive metabolite profiles of fungi might vary between closely related species or even vary between isolates of the same species. Furthermore, the activity of genes strongly depends on the cultivation conditions (e.g., the growth medium) as well as the developmental stage of a fungus (e.g., vegetative mycelium versus fruiting body).

Additionally, countless biosynthetic genes found in fungal genomes by whole-genome sequencing remain "cryptic" or "orphan" meaning their corresponding compounds are still unknown. This finding has inspired researchers to 'mine' this unearthed potential for drug discovery.[236] Bioinformatics studies on genomes of filamentous fungi portrayed a huge potential of filamentous moulds as a source of new antibiotics and other bioactive compounds.[237] These hypothetical compounds behind cryptic biosynthetic pathways can be seen as "known unknowns." Even more challenging are potential "unknown unknowns," unidentified compounds whose biosynthetic routes do not resemble characterized pathways. Identifying such compounds may take years of research, and assessing their toxicological activities requires further long-term studies. This might explain why, years after the publication of the first QPS list, the general exclusion of filamentous fungi from the QPS process remains unchanged.

3.24 Data Protection

A successful application for novel food gives rise to a generic authorisation upon which not only the original applicant but also other food business operators which fulfil the food's specifications, conditions of use and other requirements may produce the same generic product. Despite this advantage, the Novel Food Regulation

---

232 EFSA (n.d.).

233 EFSA (n.d.).

234 Brakhage et al. (2008).

235 Brakhage et al. (2008).

236 Brakhage et al. (2008).

237 Inglis et al. (2013) Nielsen et al. (2017).

provides for an effective innovation-reward balance in so far as it protects newly developed scientific evidence or scientific data supporting their applications.[238] Such data may not be used for the benefit of a subsequent application during a period of five years from the date of the authorisation of the novel food without the agreement of the initial applicant. The data protection is granted by the Commission where the following conditions are met:

(a) the newly developed scientific evidence or scientific data was designated as proprietary by the initial applicant at the time the first application was made;
(b) the initial applicant had exclusive right of reference to the proprietary scientific evidence or scientific data at the time the first application was made; and
(c) the novel food could not have been assessed by the EFSA and authorised without the submission of the proprietary scientific evidence or scientific data by the initial applicant.[239]

However, the initial applicant may agree with a subsequent applicant that such scientific evidence and scientific data may be used.

During the period of data protection, the novel food is authorised for placing on the market within the Union only by the initial applicant, unless a subsequent applicant obtains authorisation for the novel food without reference to the proprietary scientific evidence or scientific data protected or with the agreement of the initial applicant.

3.25 Examples

3.25.1 *Agaricus bisporus*[240]

(1) On 28/05/2021 EFSA received a letter from the European Commission with the request for a scientific opinion on the safety of vitamin D2 mushroom powder as a novel food.
(2) On 28/05/2021, a valid application on vitamin D2 mushroom powder as a novel food, which was submitted by the applicant, was made available to EFSA by the European Commission through the Commission e-submission portal (NF2020/2226) and the scientific evaluation procedure was initiated.
(3) On 16/07/2021, 16/05/2022 and 21/07/2023 EFSA requested the applicant to provide additional information to accompany the application and the scientific evaluation was suspended.

[238] Monaco (2023).

[239] Regulation (EU) No 2015/2283, Article 26.

[240] EFSA Panel on Nutrition, Novel Foods and Food Allergens (2024b).

(4) On 11/04/2022, 10/05/2023, 16/04/2024, additional information was provided by the applicant through the Commission e-submission portal and the scientific evaluation was restarted.
(5) During its meeting on 30/04/2024, the NDA Panel, having evaluated the data, adopted a scientific opinion on the safety of vitamin D2 mushroom powder as a NF pursuant to Regulation (EU) 2015/2283.

3.25.2 *Lentinula edodes*[241]

An example of a novel food which was granted authorisation and contains fungi is an aqueous extract obtained from the Shiitake mushroom *Lentinula edodes* cultivated in submerged fermentation. In the application for the novel food authorisation of *Lentinula edodes*, the applicant submitted that the fermentation organism used to produce the novel food, *Lentinula edodes* is commonly consumed as food and *Lentinula edodes* fruiting body is substantially equivalent to the *Lentinula edodes* mycelia used in the fermentation of the pea and rice protein. Following fermentation, there is no live *Lentinula edodes* mycelia or fungal enzymes in the final novel food preparation because of multiple heat treatment steps and thermal deactivation. The weight-of-evidence from reliable published toxicological and human clinical studies using the same or closely-related (e.g. *Lentinula edodes* mycelial extracts, reconstituted powdered *Lentinula edodes*) test materials as those components included in the novel food, support a conclusion that no adverse health effects are expected at dietary intake levels of the heat-killed *Lentinula edodes* mycelia which are estimated based on food uses proposed for the pea and rice protein fermented by Shiitake mycelia.[242]

- Compositional data on the novel food retrieved from 5 batches informs on water, pH, concentration of lentinan (the main bioactive constituent of the novel food), free glucose, N-containing constituents (using the Kjeldahl method) and proteins (using the Bradford method). The applicant further provided data on the content of fat, amino acids and related biogenic amines, ash and ions. Data on the contents of water-soluble vitamins were also provided for 3 batches. For one batch, data on heavy metals and formaldehyde was provided. Analytical methods were also used to demonstrate that contents of 297 pesticides were below the limits of detection. As for representatives of potential mycotoxins, analytical data on aflatoxins and ochratoxin was provided.
- As for secondary metabolites, the EFSA Panel agreed that under the specified fermentation conditions, the formation of secondary metabolites was unlikely. This was also supported by the history of consumption of the fruiting body: *Lentinula edodes* is indigenous to China, Japan and other Asian countries where it grows on fallen deciduous trees. It is a common food in Asia. Fresh mush-

[241] EFSA (2010b).
[242] EFSA Panel on Nutrition, Novel Foods and Food Allergens (2022c).

rooms are widely cultivated. The world production amounted to several millions of tons. The fruiting body has an established history of consumption.
- The novel food was intended to be used in a wide range of products, such as dietary supplements, yoghurts, soft drinks, cooked and processed food and baked goods.
- The applicant did not provide studies on genotoxicity. As for toxicity studies, the applicant used a literature search and provided information on studies on animals involving the acute toxicity of lentinan, a 6-month study where lentinan was given intravenously (this study was not considered relevant) and other animal studies providing no or very limited evidence to support the safety of the novel food ingredient.
- The applicant, however, conducted an efficacy and safety study of the novel food which did not provide evidence for safety concerns, although it also did not provide evidence on a possible safety margin. A second human study also provided supporting but limited evidence for the safety of the novel food.
- A Member State expressed concerns as to a potential contraindication for people suffering from auto-immune diseases. However, pointing to other studies, the Panel concluded that the risk of adverse immunological effects due to the novel food ingredient, if any, was expected not to be higher than that resulting from the normal consumption of the fruiting body of the Shiitake mushroom.
- Owing to the fermentation of the novel food ingredient from the mycelium and the final application of a heat-induced sterilisation step, adverse effects reported after the consumption of the fruiting body of the Shiitake mushroom are not considered relevant.
- There were several case reports on photosensitivity, intolerance and allergic reactions related to the consumption of Shiitake mushrooms or products derived thereof. Lentinan has been also implicated as a causative substance of Shiitake dermatitis.
- The applicant did not provide any studies to evaluate the allergenicity of the novel food. The Panel concluded that despite the low intake of lentinan at the proposed use levels, it cannot be excluded that the novel food poses an allergenic risk to sensitive subjects. However, the risk is expected not to be higher than that resulting from the normal consumption of the fruiting body of the Shiitake mushroom.

3.26 Conclusion

The novel food authorization procedure can be a long and costly process for applicants, lasting a minimum of 18 months in total and with some estimates suggesting the total cost could be over a hundred of thousands euro for a single application.[243] However, several tools are available to innovators in the authorisation process, i.e., for the pre-submission phase, preparation and submission of the application or the suitability check (see Fig. 3.6).

---

[243] European Industrial Hemp Association (2020).

Pre-submission Phase

General pre-submission advice is available to provide potential applicants or notifiers with guidance on the applicable rules and the required content for their applications prior to submission. The advice is however non-committal, non product-specific and does not include feedback on any scientific and technical work even 'indirectly relevant' to the application. Published guidance is also at risk of becoming obsolete as technologies and production techniques evolve over time.

Small and Medium Enterprises (SMEs) can request expedited pre-submission advice, which is delivered via tele-meetings. SMEs further receive tailored support in using IT tools, such as the Connect. EFSA portal and e-submission systems. This includes physical or virtual training sessions aimed at helping SMEs submit high-quality applications

Submission and Completeness/Suitability Check Phase

Clarification teleconferences are held to address any unresolved issues regarding the administrative elements of applications, or to clarify EFSA's requests for additional information during the completeness or suitability check.

Risk Assessment Phase

Clarification teleconferences are organized when EFSA requests further information during the scientific assessment, ensuring that the applicant fully understands the requirements.

Technical hearings are held to discuss additional data submitted by applicants or to address outstanding issues. These hearings may take place as part of EFSA Working Group, Panel, or Peer Review meetings. However, hearings and related requests for further information to applicants can impose significant delays to product authorisation timelines.

Post-adoption Phase

Applicants are notified of the adoption of the scientific output.

A pre-publication notification is issued for the adopted scientific output.

Post-adoption teleconferences are available to explain the scientific rationale behind EFSA's final scientific conclusions.

Services Available Throughout the Application Life Cycle

An EFSA web form, "Ask a Question," helps applicants understand general requirements for applications, procedural steps, application status, and tool usage. Follow-up support is available to clarify both administrative and scientific requirements.

Information sessions and webinars, held in-person or virtually, aim to enhance stakeholders' understanding of administrative and scientific aspects of the application process. These sessions may target specific groups, such as SMEs.

Ad hoc meetings are arranged with food and feed business representatives to discuss issues specific to certain scientific areas or challenges.

Workshops and conferences on scientific topics provide a platform for dialogue between EFSA and the broader scientific community.

Fig. 3.6 Adapted from EFSA (2021b) 'Questions and answers on the EFSA practical arrangements' [EFSA (2021b)]

Notwithstanding the availability of these tools, notably giving pre-submission advice, they have been criticised by innovators as inadequate. A more substantive, product-specific feedback would likely necessitate adopting amendments to the General Food Law.

The EFSA also offers three types of guidance documents: cross-cutting guidance on broad assessment principles, sector-specific guidance documents and other assessment methodologies produced. In 2024, the European Commission requested the EFSA to update the scientific guidance for the preparation of applications for authorisation of novel foods, previously developed following the adoption of Regulation (EU) 2015/2283 on novel foods. Recently, the EFSA has also dedicated considerable attention to alternative proteins in its scientific output, for example, as regards the allergenicity assessment of insect[244] and seaweed proteins.[245] In 2024, the EFSA published also a report on processing of novel proteins in food and feed risk assessment.[246]

Also, the EFSA has started experimenting with new risk assessment methodologies. In 2022, it finalised a project "Novel foods as red meat replacers—an insight using Risk Benefit Assessment methods" which aimed to develop and test harmonised risk benefit assessment methods.[247] The EFSA was also searching for stakeholders supporting its project "Ensuring preparedness for the assessment of safety of new food/feed sources and production technologies".[248]

3.27 Following the risk assessment

Within 7 months following the adoption of the EFSA's opinion on novel food's safety, the Commission presents a draft implementing act authorising the placing on the market of the novel food - and updating the Union List - to the Standing Committee on Plants, Animals, Food and Feed for approval via Qualified Majority Voting. Following approval of the implementing act by the Committee, the product is then included in Implementing Regulation (EU) 2017/2470—"the Union List"—which includes all authorised novel foods, and in its Annex now contains around 200 novel foods.

Post-market monitoring measures are also in place to ensure the maintenance of a high level of food safety across the value chain in the EU. Any food business operator which has placed a novel food on the market must immediately inform the Commission of any information of which it has become aware concerning (a) any new scientific or technical information which might influence the evaluation of the

[244] Liguori et al. (2022).

[245] Garciarena et al. (2022).

[246] https://efsa.onlinelibrary.wiley.com/doi/epdf/10.2903/sp.efsa.2024.EN-9113.

[247] Naska et al. (2022).

[248] EFSA (2023).

safety of use of the novel food and (b) any prohibition or restriction imposed by a third country in which the novel food is placed on the market.

3.28 Summary and Outlook for Filamentous Fungi

Taken together, the complex biology of filamentous fungi is still a major hurdle for the risk assessment. The main reasons for that are:

(1) Filamentous fungi are known to produce toxicologically relevant compounds called mycotoxins.
(2) Besides already known compounds, they have a significant genetically encoded potential to form as-yet unknown compounds derived from "silent" or "cryptic" biosynthetic gene clusters.
(3) The environmental conditions under which bioactive compounds are produced in relevant amounts remain unknown or only partially defined. Changes in cultivation conditions might even accidentally trigger the activation of silent biosynthetic gene clusters.
(4) Different developmental structures of fungi (e.g. vegetative mycelium, fruit bodies, sclerotia) often differ dramatically in their toxicological profiles.
(5) Determination of the taxonomic unit of fungi is complex. Unfortunately, relatively few genetic differences between fungal isolates, even below species level, might change the safety profile of an organism significantly.
(6) The body of knowledge about fungal bioactive compounds is still limited compared to the vast genetic potential to produce bioactive metabolites. To improve our understanding of fungi, known compounds need to be studied with regards to their toxicological effects. Furthermore, novel compounds derived from unstudied biosynthetic routes remain to be identified and toxicologically assessed.

In summary, fungal safety profiles are highly complex, which prevents regulators from granting a QPS status to filamentous fungi. This is particularly true for ascomycete molds of the subdivision *Pezizomycotina*, which are prolific mycotoxin producers and possess an exceptionally high number of biosynthetic gene clusters. Consequently, it is expected that they will continue to be handled restrictively. Mushrooms, though also known to produce toxic or harmful metabolites, generally exhibit a less extensive repertoire of biosynthetic gene clusters. Nevertheless, rarely used or less-studied mushroom species are likely to continue being treated cautiously by the EFSA for valid reasons.

However, the blanket exclusion of certain filamentous fungi from QPS may be reconsidered in the future. For example, certain well-characterized isolates of fermentation molds, such as the kōji mold, might qualify for QPS status in the EU given their longstanding GRAS status in the US. Furthermore, there is increasing evidence supporting the safety of cultivated mycelia from commonly consumed edible mushrooms. As highlighted earlier, the EFSA granted *Lentinula edodes* mycelium novel food status without requiring extensive toxicological studies. This decision may mark an initial step toward achieving QPS status for this fungus in the future.

Suggestions can be made to policy makers which could represent a further help for innovators working with novel foods derived from mushrooms and mycelium:

- The 2024 revised the EFSA Novel Food guidance is a welcome development that increases the depth and specificity of the available advice to support novel filamentous fungi producers in coming to market in the EU, including on priority areas such as allergenicity testing requirements. However, the guidance could go further in some areas, including providing clarity on the volume of tests to be carried out to demonstrate safety and providing greater differentiation between production platforms in line with best practice global examples such as the Singapore Food Agency's Requirements for the Safety Assessment of Novel Foods. The EFSA should also commit to regularly updating its guidance to avoid it becoming outdated as the pace of innovation in the sector continues.
- The EFSA could work to offer more substantive methodological support and pre-submission advice or guidance to applicants, including providing product-specific advice on test design and dossier submission.
- The EFSA could work with the European Commission to include certain fungal species on the QPS List in order to simplify safety assessments for certain well known and well characterised fungi strains.
- The EFSA could work to update the Compendium Botanicals for read-across data for fungal species.
- In the fast-paced world of food tech innovation, regulators are likely to be behind the curve of cutting-edge product development research in the field of filamentous fungi. As such, the EFSA should continue consider approvals for products in jurisdictions with similarly robust safety standards - as well as the evidence made public as part of these—as reference material when assessing applications to limit duplication and enable their risk assessment staff to focus on the most pressing food safety questions for each product. Agencies including the US Food and Drug Administration have arrangements in place with international agencies to support collaboration, while a 2023 report into the UK's Novel Food Regulation suggested the regulator could "…formally collaborate with other regulators, academia and other organisations…to share evaluation resources." The EFSA should adopt a similarly collaborative approach to ensure it does not fall behind other regulators. The precedent of EFSA's opinion on pea and rice protein fermented by shiitake (*Lentinula edodes*; synonym: *Lentinus edodes*) mycelia is promising, as it demonstrated a willingness to use regulatory findings from other markets and international expertise during assessments.

Court Cases

C-141/22 - TLL The Longevity Labs GmbH v Optimize Health Solutionsmi GmbH and BM (2023) Court of Justice of the European Union ECLI:EU:C:2023:428

C-141/22 - TLL The Longevity Labs GmbH v Optimize Health Solutionsmi GmbH and BM (2023) Opinion of Advocate General Campos Sánchez-Bordona ECLI:EU:C:2023:38

C-383/07, *M-K Europa GmbH & Co. KG v Stadt Regensburg* (2009) Court of Justice of the European Union ECLI:EU:C:2009:8

Instruments and Legislation

Austrian Federal Ministry of Social Affairs, Health, Care and Consumer Protection 'Application for consultation to determine the status of a novel food, Apricot Kernel Drink and Fermented Apricot Kernel Cream' (2021) <https://ec.europa.eu/food/system/files/2021-03/novel-food_consult-status_apricot-kernel.pdf>.

Belgium Federal Public Service of Health, Food Chain Safety and Environment, 'Application for consultation to determine the status of a novel food, pursuant to Article 4(2) of the above Regulation' (2019). https://food.ec.europa.eu/system/files/2019-02/novel-food_consult-status_lentinus-edades.pdf.

Bundesamt für Verbraucherschutz und Lebensmittelsicherheit, 'Consultation request for determination of novel food status under Article 4(2) of the above Regulation' (2021). https://food.ec.europa.eu/document/download/eaa8254a-c5cb-4864-9240-2f4159940867_en?filename=novel-food_consult-status_beverage-w-cocos.pdf .

Commission Decision 2006/721/EC of 23 October 2006 authorising the placing on the market of lycopene from Blakeslea trispora as a novel food ingredient under Regulation (EC) No 258/97 of the European Parliament and of the Council (notified under document number C(2006) 4973), OJ L 296, 26/10/2006, p. 13–16

Commission Decision 2008/968/EC of 12 December 2008 authorising the placing on the market of arachidonic acid-rich oil from Mortierella alpina as a novel food ingredient under Regulation (EC) No 258/97 of the European Parliament and of the Council (notified under document number C(2008) 8080), OJ L 344, 20.12.2008, p. 123–124

Commission Decision 2009/365/EC of 28 April 2009 authorising the placing on the market of lycopene from Blakeslea trispora as a novel food ingredient under Regulation (EC) No 258/97 of the European Parliament and of the Council (notified under document number C(2009) 3039), OJ L 111, 5.5.2009, p. 31–34

Commission Decision 2011/73/EU of 2 February 2011 authorising the placing on the market of a mycelial extract from Lentinula edodes (Shiitake mushroom) as a novel food ingredient under Regulation (EC) No 258/97 of the European Parliament and of the Council, OJ L 29, 3.2.2011, p. 30–31

Commission Decision 2011/76/EU of 2 February 2011 authorising the placing on the market of a chitin-glucan from Aspergillus niger as a novel food ingredient under Regulation (EC) No 258/97 of the European Parliament and of the Council (notified under document C(2011) 480), OJ L 29, 3.2.2011, p. 34–35

Commission Implementing Decision 2011/497/EU of 9 August 2011 authorising the placing on the market of fermented black bean extract as a novel food ingredient under Regulation (EC) No 258/97 of the European Parliament and of the Council, OJ L 205, 10.8.2011, p. 33–34

Commission Implementing Decision 2011/762/EU of 24 November 2011 authorising the placing on the market of yeast beta-glucans as a novel food ingredient under Regulation (EC) No 258/97 of the European Parliament and of the Council (notified under document C(2011) 8527), OJ L 313, 6.11.2011, p. 41–44

Commission Implementing Decision (EU) 2017/2355 of 14 December 2017 authorising the placing on the market of UV-treated mushrooms as a novel food under Regulation (EC) No 258/97 of the European Parliament and of the Council (notified under document C(2017) 8474), C/2017/8474, OJ L 336, 16.12.2017, p. 52–54

Commission Implementing Regulation (EU) 2017/2469 of 20 December 2017 laying down administrative and scientific requirements for applications referred to in Article 10 of Regulation (EU) 2015/2283 of the European Parliament and of the Council on novel foods, C/2017/8874, OJ L 351, 30.12.2017, p. 64–71

Commission Implementing Regulation (EU) 2017/2470 of 20 December 2017 establishing the Union list of novel foods in accordance with Regulation (EU) 2015/2283 of the European Parliament and of the Council on novel foods, C/2017/8878, OJ L 351, 30.12.2017, p. 72–201

Commission Implementing Regulation (EU) 2018/456 of 19 March 2018 on the procedural steps of the consultation process for determination of novel food status in accordance with Regulation (EU) 2015/2283 of the European Parliament and of the Council on novel foods, C/2018/1558, OJ L 77, 20.3.2018, p. 6–13

Czech Agriculture and Food Inspection Authority, 'Consultation request for the determination of the novel food status Article 4 of Regulation (EU) 2015/2283' (2024a). https://food.ec.europa.eu/document/download/c3301d37-d4f2-49ee-bde5-383517625b25_en?filename=novel-food_consult-status_saccharomyces-cerevisiae.pdf

Czech Agriculture and Food Inspection Authority, 'Consultation request for the determination of the novel food status Article 4 of Regulation (EU) 2015/2283' (2024b). https://food.ec.europa.eu/document/download/f652a612-6350-4ccf-92a4-dde43795db10_en?filename=novel-food_consult-status_iron-enriched-yeast.pdf

Czech Agriculture and Food Inspection Authority, 'Consultation request for the determination of the novel food status Article 4 of Regulation (EU) 2015/2283' (2024c). https://food.ec.europa.eu/document/download/4604819e-c38a-4204-91b2-f79dcbc9195e_en?filename=novel-food_consult-status_mycoprotein-fusarium-venenatum.pdf

EFSA, 'Qualified Presumption of Safety' (n.d.), https://www.efsa.europa.eu/en/applications/qps-assessment.

EFSA, 'Introduction of a Qualified Presumption of Safety (QPS) Approach for Assessment of Selected Microorganisms Referred to EFSA - Opinion of the Scientific Committee' (2007) EFSA Journal

EFSA, 'Application of Systematic Review Methodology to Food and Feed Safety Assessments to Support Decision Making' (2010a) 8 EFSA Journal

EFSA, 'Scientific Opinion on the Safety of "Lentinus Edodes Extract" (Lentinex®) as a Novel Food Ingredient' (2010b) 8 EFSA Journal 1685

EFSA, 'Scientific Opinion on the Safety of "Chitin-Glucan" as a Novel Food Ingredient' (2010c) 7 EFSA Journal 1687

EFSA, "Compendium of botanicals" (2021a). https://www.efsa.europa.eu/en/data-report/compendium-botanicals

EFSA, 'Questions and answers on the EFSA practical arrangements' (2021b). https://www.efsa.europa.eu/sites/default/files/2021-03/questions-and-answers-efsa-practical-arrangements.pdf.

EFSA, 'Calls for stakeholders' (2023). https://www.efsa.europa.eu/en/stakeholders/calls-for-stakeholders.

EFSA, 'Statement on the Requirements for Whole Genome Sequence Analysis of Microorganisms Intentionally Used in the Food Chain' (2024a) 22 EFSA Journal

EFSA, 'Updated List of QPS-Recommended Microorganisms for Safety Risk Assessments Carried out by EFSA' *Zenodo Database* (2024b) <https://zenodo.org/records/13757806>

EFSA Panel on Additives and Products or Substances used in Animal Feed (FEEDAP), 'Guidance on the Assessment of Bacterial Susceptibility to Antimicrobials of Human and Veterinary Importance' (2012) 10 EFSA Journal

EFSA Panel on Biological Hazards (BIOHAZ), 'Update of the List of QPS-recommended Biological Agents Intentionally Added to Food or Feed as Notified to EFSA 12: Suitability of Taxonomic Units Notified to EFSA until March 2020' (2020) 18 EFSA Journal

EFSA Panel on Biological Hazards, 'Update of the List of QPS-Recommended Biological Agents Intentionally Added to Food or Feed as Notified to EFSA 15: Suitability of Taxonomic Units Notified to EFSA until September 2021' (2022) 20 EFSA Journal e07045

EFSA Panel on Biological Hazards (BIOHAZ), 'Update of the List of Qualified Presumption of Safety (QPS) Recommended Microorganisms Intentionally Added to Food or Feed as Notified to EFSA' (2023) 21 EFSA Journal

EFSA Panel on Biological Hazards (BIOHAZ), 'Update of the List of Qualified Presumption of Safety (QPS) Recommended Microbiological Agents Intentionally Added to Food or Feed as Notified to EFSA 19: Suitability of Taxonomic Units Notified to EFSA until September 2023' (2024) 22 EFSA Journal e8517

EFSA Panel on Dietetic Products, Nutrition and Allergies (NDA), 'Guidance on the Preparation and Presentation of Applications Pursuant to Article 21 Paragraph 2 of Regulation (EU) No 1169/2011' (2013) 11 EFSA Journal

EFSA Panel on Dietetic Products, Nutrition and Allergies (NDA), 'Scientific Opinion on the Safety of Vitamin D-enriched UV-treated Baker's Yeast' (2014) 12 EFSA Journal

EFSA Panel on Food Contact Materials, Enzymes and Processing Aids (EFSA CEP Panel) and others, 'Scientific Guidance for the Submission of Dossiers on Food Enzymes' (2021) 19 EFSA journal. European Food Safety Authority e06851

EFSA Panel on Nutrition, Novel Foods and Food Allergens (NDA),'Safety of Yarrowia Lipolytica Yeast Biomass as a Novel Food Pursuant to Regulation (EU) 2015/2283' (2019) 17 EFSA Journal

EFSA Panel on Nutrition, Novel Foods and Food Allergens (NDA), 'Safety of Vitamin D2 Mushroom Powder as a Novel Food Pursuant to Regulation (EU) 2015/2283' (2020a) 18 EFSA Journal e5948

EFSA Panel on Nutrition, Novel Foods and Food Allergens (NDA), 'Safety of Chromium-enriched Biomass of Yarrowia Lipolytica as a Novel Food Pursuant to Regulation (EU) 2015/2283' (2020b) 18 EFSA Journal

EFSA Panel on Nutrition, Novel Foods and Food Allergens (NDA), 'Safety of lacto-N-neotetraose (LNnT) Produced by Derivative Strains of E. Coli BL21 as a Novel Food Pursuant to Regulation (EU) 2015/2283' (2020c) 18 EFSA Journal

EFSA Panel on Nutrition, Novel Foods and Food Allergens (NDA), 'Safety of Vitamin D2 Mushroom Powder (Agaricus Bisporus) as a Novel Food Pursuant to Regulation (EU) 2015/2283' (2021a) 19 EFSA Journal

EFSA Panel on Nutrition, Novel Foods, Food Allergens (NDA), 'Safety of Extended Uses of UV-treated Baker's Yeast as a Novel Food Pursuant to Regulation (EU) 2015/2283' (2021b) 19 EFSA Journal

EFSA Panel on Nutrition, Novel Foods and Food Allergens (NDA), 'Safety of Dried Yellow Mealworm (Tenebrio Molitor Larva) as a Novel Food Pursuant to Regulation (EU) 2015/2283' (2021c) 19 EFSA Journal

EFSA Panel on Nutrition, Novel Foods and Food Allergens (NDA), 'Safety of Freeze-dried Mycelia of *Antrodia Camphorata* as a Novel Food Pursuant to Regulation (EU) 2015/2283' (2022a) 20 EFSA Journal

EFSA Panel on Nutrition, Novel Foods and Food Allergens (NDA), 'Safety of vitamin D2 mushroom powder as a Novel food pursuant to Regulation (EU) 2015/2283 (NF 2019/1471)' (2022b) 20 EFSA Journal

EFSA Panel on Nutrition, Novel Foods and Food Allergens (NDA), 'Safety of Pea and Rice Protein Fermented by Shiitake (*Lentinula Edodes*) Mycelia as a Novel Food Pursuant to Regulation (EU) 2015/2283' (2022c) 20 EFSA Journal

EFSA Panel on Dietetic Products, Nutrition and Allergies (NDA), 'Guidance on the Scientific Requirements for an Application for Authorisation of a Novel Food in the Context of Regulation (EU) 2015/2283' (2024a) 22 EFSA Journal e8961

EFSA Panel on Dietetic Products, Nutrition and Allergies (NDA), 'Safety of Vitamin D2 Mushroom Powder as a Novel Food Pursuant to Regulation (EU) 2015/2283 (NF 2020/2226)' (2024b) 22 EFSA Journal

EFSA Scientific Committee, 'Guidance on Selected Default Values to Be Used by the EFSA Scientific Committee, Scientific Panels and Units in the Absence of Actual Measured Data' (2012) 10 EFSA Journal

EFSA Scientific Committee, 'Evaluation of Existing Guidelines for Their Adequacy for the Food and Feed Risk Assessment of Microorganisms Obtained through Synthetic Biology' (2022) 20 EFSA Journal

European Commission, 'Guidance on the interpretation of key provisions of Directive 2008/98/EC on waste' (2008) http://waste-prevention.gr/waste/wp-content/uploads/2015/10/2012_Guidance%20interpretation%20Directive%2098-2008-EC_EN.pdf.

European Commission, 'Human Consumption to a Significant Degree Information and Guidance Document' (2016), https://food.ec.europa.eu/system/files_en?file=2016-10/novel-food_guidance_human-consumption_en.pdf.

European Commission, 'Summary of the application: Fungi protein from *Fusarium* strain *flavolapis*' (2021). https://food.ec.europa.eu/system/files/2022-11/novel-food_sum_ongoing-app_2021-1350.pdf.

European Commission, 'Consultation process on novel food status' (2023). https://food.ec.europa.eu/safety/novel-food/consultation-process-novel-food-status_en.

European Commission, 'Novel Food Catalogue' (2024a), https://food.ec.europa.eu/food-safety/novel-food/novel-food-status-catalogue_en

European Commission 'Summary of the application: mycelium from oyster mushroom' (2024b) https://food.ec.europa.eu/document/download/7530c063-628a-43b6-8100-5520a81d04e3_en?filename=novel-food_sum_ongoing-not_2023-15001.pdf

European Parliament, 'At glance: Speeding up authorisation of novel foods' (2015) https://www.europarl.europa.eu/RegData/etudes/ATAG/2015/569022/EPRS_ATA(2015)569022_EN.pdf.

FAO, 'Development of a Protein Database and the Way Forward for Reviewing Protein Requirements' (2022) <https://www.fao.org/family-farming/detail/en/c/1698270/>

Federal Ministry of Social Affairs, Health, Care and Consumer Protection, 'Application for consultation to determinate the status of a novel food, pursuant to Article 4(2) of the Regulation (EU) 2015/2283 of the European Parliament and of the Council of 25 November 2015 on novel foods' (2022). https://food.ec.europa.eu/system/files/2022-06/novel-food_consult-status_flour-buckwheat.pdf

Finnish Food Authority, 'Application for consultation to determine the status of a novel food pursuant to Article 4(2) of the Regulation (EU) 2015/2283 of the European Parliament and of the Council of 25 November 2015 on novel foods' (2022a). https://food.ec.europa.eu/system/files/2022-06/novel-food_consult-status_4204527_0.pdf.

Finnish Food Authority, 'Application for consultation to determine the status of a novel food pursuant to Article 4(2) of the Regulation (EU) 2015/2283 of the European Parliament and of the Council of 25 November 2015 on novel foods' (2022b).: https://food.ec.europa.eu/system/files/2022-09/novel-food_consult-status_paecilomyces-variotii.pdf

Finnish Food Authority, 'Application for consultation to determine the status of a novel food pursuant to Article 4(2) of the Regulation (EU) 2015/2283 of the European Parliament and of the Council of 25 November 2015 on novel foods'

(2024), https://food.ec.europa.eu/document/download/bc4d6da9-87ae-4198-8bde-78ea71d3863f_en?filename=novel-food_consult-status_chaga_fin.pdf.
Food Safety Authority of Ireland, 'Application for the approval of UV-treated mushrooms with increased vitamin D levels under the novel food Regulation (EC) No. 258/97' (2016). https://food.ec.europa.eu/system/files/2016-10/novel-food_authorisation_2016_auth-letter_uv-mushrooms_en.pdf.
Food Safety Authority of Ireland, 'Application for the approval of UV-treated mushrooms (*Agaricus Bisporus*) with increased vitamin D levels under the novel food Regulation (EC) No. 258/97' (2017). https://food.ec.europa.eu/system/files/2017-09/novel-food_authorisation_2017_auth-letter_uv-mushrooms_en.pdf.
Food Safety Authority of Ireland, 'Article 4 Request Regulation (EU) 2015/2283' (2019). https://food.ec.europa.eu/system/files/2019-01/novel-food_consult-status_aspergillus-oryzae.pdf.
Food Safety Authority of Ireland, 'Article 4 Request Regulation (EU) 2015/2283' (2022a). https://food.ec.europa.eu/system/files/2022-01/novel-food_consult-status_fsai-dried-yeast-fermantate.pdf.
Food Safety Authority of Ireland, 'Article 4 Request Regulation (EU) 2015/2283' (2022b). https://food.ec.europa.eu/system/files/2022-01/novel-food_consult-status_torula-yeast.pdf.
Food Safety Authority of Ireland, 'Article 4 Request Regulation (EU) 2015/2283' (2023). https://food.ec.europa.eu/system/files/2023-01/novel-food_consult-status_se-vit-b12-mushrooms.pdf
Food Standards Agency, 'Application for consultation to determine the status of a novel food, pursuant to Article 4(2) of the above Regulation' (2019). https://food.ec.europa.eu/system/files/2019-12/novel-food_consult-status_agaricus-bisporus.pdf.
Food Standards Agency, 'Consultation request to determine novel food status, pursuant to Article 4(2) of Retained Regulation (EU) 2015/2283' (2022). https://www.foodstandards.gov.scot/downloads/Article_4_consultation_request_-_Monaghan_Mushrooms_-_B12_and_Selenium_Enhanced_mushrooms_-_decision_letter_-_13_June_2022.pdf.
Italian Ministry of Health, 'Application for consultation to determine the status of a novel food, pursuant to Article 4(2) of the Regulation (UE) 2015/2283' (2022). https://food.ec.europa.eu/system/files/2022-04/novel-food_consult-status_whey-treated-uv-rad.pdf.
Livsmedelsverket, 'Consultation request for the determination of novel food status pursuant to Article 4(2) of the Regulation (EU) 2015/2283 of the European Parliament and of the Council of 25 November 2015 on novel foods' (2022). https://food.ec.europa.eu/system/files/2022-06/novel-food_consult-status_4515555.pdf.
National Food Chain Safety Office of Hungary, 'Consultation request for a determination of novel food status pursuant to Article 4(2) of the above Regulation' (2020). https://food.ec.europa.eu/system/files/2020-04/novel-food_consult-status_fermented-wheat.pdf.

Protocol on Ireland/Northern Ireland to the Agreement on the withdrawal of the United Kingdom of Great Britain and Northern Ireland from the European Union and the European Atomic Energy Community. 2019/C 384 I/01.

Regulation (EC) No 852/2004 of the European Parliament and of the Council of 29 April 2004 on the hygiene of foodstuffs, OJ L 139, 30.4.2004, p. 1–54

Regulation (EC) No 853/2004 of the European Parliament and of the Council of 29 April 2004 laying down specific hygiene rules for food of animal origin, OJ L 139, 30.4.2004, p. 55–205

Regulation (EU) 2015/2283 of the European Parliament and of the Council of 25 November 2015 on novel foods, amending Regulation (EU) No 1169/2011 of the European Parliament and of the Council and repealing Regulation (EC) No 258/97 of the European Parliament and of the Council and Commission Regulation (EC) No 1852/2001, OJ L 327, 11.12.2015, p. 1–22

Regulation (EU) 2019/1381 of the European Parliament and of the Council of 20 June 2019 on the transparency and sustainability of the EU risk assessment in the food chain and amending Regulations (EC) No 178/2002, (EC) No 1829/2003, (EC) No 1831/2003, (EC) No 2065/2003, (EC) No 1935/2004, (EC) No 1331/2008, (EC) No 1107/2009, (EU) 2015/2283 and Directive 2001/18/EC, PE/41/2019/REV/1, OJ L 231, 6.9.2019, p. 1–28

Regulation (EU) No 1169/2011 of the European Parliament and of the Council of 25 October 2011 on the provision of food information to consumers, amending Regulations (EC) No 1924/2006 and (EC) No 1925/2006 of the European Parliament and of the Council, and repealing Commission Directive 87/250/EEC, Council Directive 90/496/EEC, Commission Directive 1999/10/EC, Directive 2000/13/EC of the European Parliament and of the Council, Commission Directives 2002/67/EC and 2008/5/EC and Commission Regulation (EC) No 608/2004, OJ L 304, 22.11.2011, p. 18–63

Výnos Ministerstva pôdohospodárstva Slovenskej republiky a Ministerstva zdravotníctva Slovenskej republiky zo 7. júna 2000 c. 1541/3/2000 -100, ktorým sa vydáva hlava Potravinového kódexu Slovenskej republiky upravujúca jedlé huby a výrobky z jedlých húb.

Slovak Food Codex (2000) https://www.nahuby.sk/clanok.php?clanok_id=19

Spanish Agency for Food Safety and Nutrition, 'Application for consultation to determine the status of a novel food, pursuant to Article 4(2) of the Regulation (EU) 2015/2283 of the European Parliament and of the Council of 25 November 2015 on novel foods' (2019a). https://food.ec.europa.eu/system/files/2019-10/novel-food_consult-status_lentinula-edodes_aesan.pdf.

Spanish Agency for Food Safety and Nutrition, 'Application for consultation to determine the status of a novel food, pursuant to Article 4(2) of the Regulation (EU) 2015/2283 of the European Parliament and of the Council of 25 November 2015 on novel foods' (2019b). https://food.ec.europa.eu/system/files/2019-10/novel-food_consult-status_agaricus-blazei_aesan.pdf

Spanish Agency for Food Safety and Nutrition, 'Application for consultation to determine the status of a novel food, pursuant to Article 4(2) of the Regulation (EU) 2015/2283 of the European Parliament and of the Council of 25 November 2015 on novel foods' (2019c). https://food.ec.europa.eu/system/files/2019-10/novel-food_consult-status_coprinus-comatus_aesan.pdf

Spanish Agency for Food Safety and Nutrition, 'Application for consultation to determine the status of a novel food, pursuant to Article 4(2) of the Regulation (EU) 2015/2283 of the European Parliament and of the Council of 25 November 2015 on novel foods' (2019d). https://food.ec.europa.eu/system/files/2019-10/novel-food_consult-status_pleorotus-ostreatus_aesan.pdf

Spanish Agency for Food Safety and Nutrition, 'Application for consultation to determine the status of a novel food, pursuant to Article 4(2) of the Regulation (EU) 2015/2283 of the European Parliament and of the Council of 25 November 2015 on novel foods' (2019e). https://food.ec.europa.eu/system/files/2019-10/novel-food_consult-status_pleorotus-eryngii_aesan.pdf

Spanish Agency for Food Safety and Nutrition, 'Application for consultation to determine the status of a novel food, pursuant to Article 4(2) of the Regulation (EU) 2015/2283 of the European Parliament and of the Council of 25 November 2015 on novel foods' (2019f). https://food.ec.europa.eu/system/files/2019-10/novel-food_consult-status_polyporus-umbellatus_aesan.pdf

Spanish Agency for Food Safety and Nutrition, 'Application for consultation to determine the status of a novel food, pursuant to Article 4(2) of the Regulation (EU) 2015/2283 of the European Parliament and of the Council of 25 November 2015 on novel foods' (2019g). https://food.ec.europa.eu/system/files/2019-10/novel-food_consult-status_grifola-frondosa_aesan.pdf

Spanish Agency for Food Safety and Nutrition, 'Application for consultation to determine the status of a novel food, pursuant to Article 4(2) of the Regulation (EU) 2015/2283 of the European Parliament and of the Council of 25 November 2015 on novel foods' (2019h). https://food.ec.europa.eu/system/files/2019-10/novel-food_consult-status_hericium-erinaceus_aesan.pdf

Spanish Agency for Food Safety and Nutrition 'Application for consultation to determine the status of a novel food, pursuant to Article 4(2) of the Regulation (EU) 2015/2283 of the European Parliament and of the Council of 25 November 2015 on novel foods' (2019i). https://food.ec.europa.eu/system/files/2019-10/novel-food_consult-status_ganoderma-lucidum_aesan.pdf

Spanish Agency for Food Safety and Nutrition, 'Application for consultation to determine the status of a novel food, pursuant to Article 4(2) of the Regulation (EU) 2015/2283 of the European Parliament and of the Council of 25 November 2015 on novel foods' (2023). https://food.ec.europa.eu/system/files/2023-01/novel-food_consult-status_2023-306385.pdf.

References

Bibliography

Barzee TJ et al (2021) Fungi for future foods. J Fut Foods 1:25

Boukid F et al (2023) Fermentation for designing innovative plant-based meat and dairy alternatives. Foods 12:1005

Brakhage AA et al. (2008) Activation of fungal silent gene clusters: a new avenue to drug discovery. Progress in Drug Research. Fortschritte Der Arzneimittelforschung. Progres Des Recherches Pharmaceutiques 66:1

Chang P-K (2019) Genome-wide Nucleotide Variation Distinguishes *Aspergillus Flavus* from *Aspergillus Oryzae* and Helps to Reveal Origins of Atoxigenic *A. Flavus* Biocontrol Strains. J Appl Microbiol 127:1511

Constable A et al. (2007) History of safe use as applied to the safety assessment of novel foods and foods derived from genetically modified organisms. Food Chem Toxicol 45:2513

Coton E et al (2020) Penicillium Roqueforti: an overview of its genetics, physiology, metabolism and biotechnological applications. Fungal Biol Rev 34:59

Daba GM, Mostafa FA, Elkhateeb WA (2021) The Ancient Koji Mold (Aspergillus Oryzae) as a modern biotechnological tool. Bioresour Bioproc 8:52

de Boer A, Bast A (2018) Demanding safe foods – safety testing under the novel food regulation (2015/2283). Trends Food Sci Technol 72:125

Directive 2004/10/EC of the European Parliament and of the Council of 11 February 2004 on the harmonisation of laws, regulations and administrative provisions relating to the application of the principles of good laboratory practice and the verification of their applications for tests on chemical substances, OJ L 50, 20.2.2004, pp 44–59

Directive 2008/98/EC of the European Parliament and of the Council of 19 November 2008 on waste and repealing certain Directives, OJ L 312, 22.11.2008, pp 3–30

Dongo D, Della Penna A (2023) Novel foods, the state of the art in the European Union. Food Times, 21 November 2023, https://www.foodtimes.eu/research-en/novel-foods-the-state-of-the-art-in-the-european-union/

Dörfer M, Gressler M, Hoffmeister D (2019) Diversity and bioactivity of Armillaria Sesquiterpene Aryl Ester natural products. Mycol Progr 18:1027

Dowlati M et al (2021) Heavy metals content in edible mushrooms: a systematic review, meta-analysis and health risk assessment. Trends Food Sci Technol 109:527

Dufossé L et al (2014) Filamentous fungi are large-scale producers of pigments and colorants for the food industry. Curr Opin Biotechnol 26:56

Edelman J, Fewell A, Solomons G (1983) Myco-protein – a new food. Nutr Abstr Rev Clin Nutr 53:471

European Industrial Hemp Association (2020) EIHA Novel Food Consortium. https://eiha.org/wp-content/uploads/2020/09/Novel-Food-info-text.pdf

Food Supplements Europe (2019) Guidance for food business operators on the verification of the status of a new food under the new Novel Foods. https://foodsupplementseurope.org/wp-content/themes/fse-theme/documents/publications-and-guidelines/novelfoods-guidelines-jan2019.pdf

Frisvad JC et al (2007) Fumonisin B_2 production by Aspergillus Niger. J Agric Food Chem 55:9727

Garciarena IN, Poulsen M, Lindholm Bøgh K (2022) Risk–benefit assessment of seaweed allergenicity risk assessment of novel protein. EFSA J 20

Good Food Institute (2024) The Science of Fermentation. https://gfi.org/science/the-science-of-fermentation/

Hadi J, Brightwell G (2021) Safety of alternative proteins: technological, environmental and regulatory aspects of cultured meat, plant-based meat, insect protein and single-cell protein. Foods 10:1226

Hawksworth DL et al (2011) The Amsterdam declaration on fungal nomenclature. IMA Fungus 2:105

Heinonen M, Mangelsdorf I (2017) Webinar on Scientific Aspects to consider when preparing a Novel Food Application. https://www.efsa.europa.eu/sites/default/files/170615-p01.pdf

Ho S (2021) Spanish startup libre foods wants to be "World's Leading Provider" of Whole-Cut Mycelium Steaks. <https://www.greenqueen.com.hk/spanish-startup-libre-foods-wants-to-be-worlds-leading-provider-of-whole-cut-mycelium-steaks/>

Inglis DO et al. (2013) Comprehensive annotation of secondary metabolite biosynthetic genes and gene clusters of Aspergillus Nidulans, A. Fumigatus, A. Niger and A. Oryzae. BMC Microbiol 13:91

Jørgensen TR (2007) Identification and toxigenic potential of the industrially important fungi, Aspergillus Oryzae and Aspergillus Sojae. J Food Prot 70:2916

Katona SJ (2002) Sensitivity to Quorn Mycoprotein (Fusarium Venenatum) in a Mould allergic patient. J Clin Pathol 55:876

Kensler TW et al. (2011) Aflatoxin: a 50-year Odyssey of mechanistic and translational toxicology. Toxicol Sci 120:S28

Kirsch LS, de Macedo AJP, Teixeira MFS (2016) Production of Mycelial Biomass by the Amazonian Edible Mushroom Pleurotus Albidus. Braz J Microbiol 47:658

Kurtzman CP et al (1986) Dna relatedness among wild and domesticated species in the *Aspergillus Flavus* group. Mycologia 78:955

Kurtzman CP, Fell JW, Boekhout T (2011) 'Chapter 1 - definition, classification and nomenclature of the yeasts. In: Kurtzman CP, Fell JW, Boekhout T (eds) The yeasts. Elsevier

Li YF et al (2016) Comprehensive curation and analysis of fungal biosynthetic gene clusters of published natural products. Fungal Genet Biol 89:18

Liguori B et al (2022) Novel foods: allergenicity assessment of insect proteins. EFSA J 20

López Sánchez de León L, et al. (2023) Novel foods authorised in the European Union: analysis and assessment of their specifications. https://www.aesan.gob.es/AECOSAN/docs/documentos/seguridad_alimentaria/evaluacion_riesgos/informes_cc_ingles/COLLABORATION_NOVEL_FOODS_EU.pdf

May TW et al. (2019) Chapter F of the International Code of Nomenclature for Algae, Fungi, and Plants as Approved by the 11th International Mycological Congress, San Juan, Puerto Rico, July 2018. IMA Fungus 10:21

Ministry of Health, Welfare and Sport of the Netherlands (2016) Arachidonic acid-rich oil from the fungus Mortierella alpine. https://food.ec.europa.eu/system/files/2016-10/novel-food_authorisation_2011_auth-letter_arachidonic_acid_rich_oil_en.pdf

Molitorisová A, Monaco A, Purnhagen K (2021) An analysis of the regulatory framework applicable to products obtained from mushroom and mycelium

Monaco A (2023) Data protection under the novel food regulation: valuable instrument or barrier to innovation? Insights from the insect sector. Eur Food Feed Law Rev 18

Naranjo-Ortiz MA, Gabaldón T (2019) Fungal evolution: diversity, taxonomy and phylogeny of the fungi. Biol Rev 94:2101

Naska A et al (2022) Novel foods as red meat replacers – an insight using risk benefit assessment methods (the NovRBA Project). EFSA Supporting Publications, p 19

Nielsen JC et al (2017) Global analysis of biosynthetic gene clusters reveals vast potential of secondary metabolite production in Penicillium species. Nat Microbiol 2:17044

Pariza MW et al (2015) Determining the safety of microbial cultures for consumption by humans and animals. Regul Toxicol Pharmacol 73:164

Patriarca A, Fernández Pinto V (2017) Prevalence of Mycotoxins in foods and decontamination. Curr Opinion Food Sci 14:50

Richard JL (2007) Some major Mycotoxins and their Mycotoxicoses—an overview. Int J Food Microbiol 119:3

Ronchetti F, Springer L, Purnhagen K (2024) The regulatory landscape in the EU for dairy products derived from precision fermentation. Springer Briefs in Law

Rzymski P et al (2017) Cultivation of *Agaricus Bisporus* Enriched with Selenium, Zinc and Copper. J Sci Food Agric 97:923
Scott PM (1981) Toxins of Penicillium species used in cheese manufacture. J Food Protect 44:702
Sevik H et al (2020) Changes in heavy metal accumulation in some edible landscape plants depending on traffic density. Environ Monit Assessment 192:78
Stankovic S et al (2012) Heavy metals in seafood Mussels. Risks for human health. In: Lichtfouse E, Schwarzbauer J, Robert D (eds) Environmental chemistry for a sustainable world. Springer, Netherlands
Tian F et al (2022) Mycotoxins in Soybean-based foods fermented with Filamentous Fungi: occurrence and preventive strategies. Comprehensive Rev Food Sci Food Saf 21:5131
Tominaga M et al (2006) Molecular analysis of an inactive Aflatoxin biosynthesis gene cluster in *Aspergillus Oryzae* RIB strains. Appl Environ Microbiol 72:484
Turck D et al (2016) Guidance on the preparation and presentation of an application for authorisation of a novel food in the context of regulation (EU) 2015/2283. EFSA J 14:e04594
van der Fels-Klerx HJ et al (2024) Framework for evaluation of food safety in the circular food system. NPJ Sci Food 8:36
Van der Meulen BMJ (2013) The structure of European food law. Laws 2:69
Waldbillig A et al (2023) Exploring Psilocybe Spp. Mycelium and fruiting body chemistry for potential therapeutic compounds. Front Fungal Biol 4:1295223
Wiebe M (2002) Myco-Protein from Fusarium Venenatum: a well-established product for human consumption. Appl Microbiol Biotechnol 58:421
Wikandari R et al (2023) Filamentous fungi for food. Curr Dev Biotechnol Bioeng: Filamentous Fungi Biorefinery 353
Zwinkels J, Wolkers-Rooijackers J, Smid EJ (2023) Solid-state fungal fermentation transforms low-quality plant-based foods into products with improved protein quality. LWT 184:114979

Chapter 4
Labelling of Mushrooms and Mycelium Products

Abstract This chapter provides a comprehensive guide on the key information that must be communicated to consumers regarding mushroom and mycelium-based products. It begins by examining the specific naming conventions applicable to mycelium products. The focus then shifts to the regulations governing claims made on these products, particularly the use of "plant-based" claims as well as health and nutrition claims associated with mushrooms and mycelium-based items. The chapter further offers an overview of additional voluntary information that food businesses may choose to provide. This includes an exploration of the concept of clean labelling, criteria for vegan and vegetarian designations, requirements for organic certification, and a discussion of recent regulatory developments concerning the use of environmental claims in food products.

4.1 Labelling Requirements for MMP

'Food information' is defined in Regulation (EU) No 1169/2011 on the provision of food information to consumers (FIC) as any information concerning a food made available to consumers by means of a label or any other means, including modern technology tools or verbal communication.[1] Such a broad definition implies that even social media posts or website statements made by food business operators must adhere to the general principles of the FIC.

Food information can either be mandated by law or voluntarily provided by food business operators. In both cases, food information must not be misleading regarding the characteristics of the food and its properties and is meant to be accurate, clear, and easy to understand.[2] When evaluating whether information provided to consumers is misleading, the "average consumer" benchmark is normally applied. Derived from CJEU case law, the definition of average consumers presumes that a

[1] Regulation (EU) No 1169/2011, Article 2(2)(a).

[2] Regulation (EU) No 1169/2011, Article 7.

A. Molitorisová et al., *Regulation of Mushroom- and Mycelium-Based Products in the EU*, SpringerBriefs in Law, https://doi.org/10.1007/978-3-031-93780-4_4

consumer is reasonably well informed and reasonably observant and circumspect.[3] This means that in principle in the EU, consumers are expected to read the label and understand it, considering the list of ingredients. However, even a precise list of ingredients is not considered enough to correct an erroneous impression generated by the overall appearance of the products, including the label.[4]

The FIC distinguishes between regulations concerning mandatory and voluntary information. Among the mandatory information are:[5] the name of the food; the list of ingredients; the presence of allergens;[6] the quantity of certain ingredients and the categories of ingredients; the net quantity of food; the date of minimum durability and the "use by" date; special storage conditions and/or the condition of use; the name of the business name and the address of the food business operator responsible for the food information; for certain foods, the country of origin or place of provenance; and the nutrition declaration. Among these, the name of the food is of particular interest when discussing novel foods.

Regarding voluntary information, several claims are normally associated with MMP, in particular claims related to their status as "plant-based" meat alternatives, health and nutrition claims, vegan and vegetarian, organic and sustainability claims. These claims are subject to differing rules and provisions at the EU and national level and need to be based on reliable information and data.

4.2 The Legal Name of Novel Foods Obtained from Mycelia

Novel food applicants are required to suggest which information should appear on the novel food label to not mislead consumers or a verifiable justification for why those elements are not necessary.[7] The labeling conditions are assessed during the authorization procedure and are subsequently specified in the Union List of novel foods. Any food business operators wishing to place these novel foods on the market must comply with these conditions. Such labelling specifications inform the consumer of the characteristic of the novel food that renders it no longer equivalent to an existing food, keeping into consideration its composition, nutritional value and potential implication for specific groups of the population.[8] Additionally, selecting

[3] *C-210/96* (1998), paragraph 37; *C-470/93* (1995), paragraph 24.

[4] *C-195/14* (2015), paragraphs 39 to 44.

[5] Regulation (EU) No 1169/2011, Article 9.

[6] Regulation (EU) No 1169/2011, Annex II. The Annex lists 14 allergens that must be indicated in the label of a food products. Mushrooms are not in the list. In general, companies working on MMP tend to claim an absence of allergens in their products. It is always wise to support such claims with appropriate studies. Although not being considered of high-level risks, allergic reactions from mushrooms might still occur. If a substantial risk is identified during the novel food application, labelling might be required among the Conditions of Use.

[7] Regulation (EU) 2015/2283, Article 10(2)(g).

[8] Regulation (EU) 2015/2283, Article 9(3)(b).

an appropriate name for novel foods will be critical in informing consumers about the product's nature and influencing its acceptance.[9]

Every foodstuff sold in the EU must bear a legal name. The legal name of the food is not the marketing name of the product. The marketing name of the product is a marketing tool, and it is the decision of the food business operator selling the food, whether to include it or not, to attract consumers.[10] The legal name of the food product instead provides information on the actual nature of the product and must be clearly displayed on the label. The legal name must be either its name as prescribed by the law (e.g. the name 'milk chocolate' is regulated at the EU level and only certain ingredients can be used for producing it[11]), a customary name, comprehensible to the consumers without being misleading (e.g. Caesar salad), or a descriptive name (e.g. baked bread with ham and cheese).[12] If a certain food is then included as an ingredient in another food product, the legal name must be used in the list of ingredients of the latter. Since they are normally classified as novel foods, products obtained from mycelium will all have their legal name specified in the authorization, based on the suggestions made by the applicants.

Currently, there are five authorized novel foods obtained from fungal mycelium: chitin-glucan from *Aspergillus niger*, chitosan extract from *Aspergillus niger* (obtained from the mycelium of *Aspergillus niger*), the mycelial extract from Shiitake mushroom (*Lentinula edodes*), the pea and rice protein fermented by *Lentinula edodes* (Shiitake mushroom) mycelia, and the *Antrodia camphorata* mycelia powder.[13]

While for the "Pea and rice protein fermented by Shiitake mushroom mycelia" and for the "*Antrodia camphorata* mycelia powder" the labelling requirements clarify the mycelium origin of the novel foods, this is not the case for the other mycelium-based novel foods. Designations as: 'extract from the mushroom Lentinula edodes' or 'extract from Shiitake mushroom' are considered enough.

Since the novel food status of these products obtained from commonly consumed fungi is due to the use of the mycelia instead of the fruiting bodies, we are of the opinion that the mycelium origin should always be specified to avoid misleading consumers, as for the two more recent authorizations.

[9] Hallman and Hallman (2020).

[10] Charette et al. (2015).

[11] Directive 2000/36/EC, Annex I.

[12] Regulation (EU) No 1169/2011, Article 2(2)(m)(n)(o) and Article 17.

[13] Commission Implementing Regulation (EU) 2017/2470, Annex, Table 1.

4.3 Mycelium of … Generic Name or Scientific Species?

Considering that commonly consumed mushrooms are labelled with the scientific name of the species or the generic name under which consumers know them, mycelium products should follow the same logic, by considering how the general public typically identifies the various species.

For the novel food "*Antrodia camphorata* mycelia powder" the scientific name is used in the labelling indications. This is in line with other novel foods, like for example the biomass obtained from the protozoa *Euglena gracilis* (whose required labelling is 'dried biomass of *Euglena gracilis* algae') and for three different biomass products from the yeast *Yarrowia lipolytica,* labelled respectively as 'chromium-containing yeast (*Yarrowia lipolytica*) biomass', '*Yarrowia lipolytica* yeast heat-killed biomass' and 'selenium-containing yeast (*Yarrowia lipolytica*) biomass'.

On the contrary, the labelling conditions of "Pea and rice protein fermented by *Lentinula edodes*" only require the generic name of the species ("Pea and rice protein fermented by Shiitake mushroom mycelia"). This is due to the widespread use of the term "Shiitake mushrooms," which takes precedence over the use of the scientific name.

For other MMP, the use of both the scientific and the generic names might be more appropriate. This would be in line with other authorized novel foods. For example, the "Antarctic Krill oil" is labelled as 'Lipid extract from the crustacean Antarctic Krill (*Euphausia superba*),' while chia products are labelled as 'Chia oil (*Salvia hispanica*)' and 'Chia seeds (*Salvia hispanica*).'

4.4 "Mycoproteins" and Similar Designations

Several companies working with protein extracts from MMP have showed interest in using the term "mycoproteins" or similar wordings, in the marketing materials of their products. As discussed, when foodstuffs are authorized as novel foods their specific labelling requirements are specified in the authorization's conditions. Generic terms like mycoproteins are not likely to be authorized as the sole labelling designation for MMP classified as novel foods because they are too generic. However, they can be used for marketing purposes.

Mycoprotein is already used by at least one company present on the market for referring to a product obtained from *Fusarium venenatum.*[14] Mycoprotein and similar single-word formulations might confuse consumers, particularly when referring to organism part of the kingdom fungi but with a different composition than commonly consumed mushrooms. To avoid any confusion, a disclaimer like 'protein extracted from mycelium biomass of …' or 'proteins from … (generic name and/or

[14] Brennan (2021).

scientific name of the species) mycelium' should always accompany "mycoproteins" or similar trademark names in the marketing materials and websites.

In the United States, the use of the wording 'mycoprotein' for labelling a wide range of fungi-derived products raised wide public attention when it was linked to the death of an 11-years old boy in California. The child ate a vegan version of chicken nuggets, containing 'mycoprotein' obtained from *Fusarium venenatum*, a mould part of the fungi family. The family of the victim did not know that the product contained mould (to which the child was allergic); they believed the product was simply mushroom-based. The company denied all allegations and claimed that the child died because of an asthma attack, referring to the autopsy results. They also pointed out to a vast number of products containing *Fusarium venenatum* sold over the years. Responding to increasing public concerns, the company agreed to indicate on the label that the product in question contains a mould.[15]

In the EU, Annex II of the FIC lists 14 allergens that must be indicated on the label of food products. Fungi and mushrooms are not present on this list. Companies developing MMP sometimes claim that their products are free from allergens.[16] When making such claims, conducting appropriate allergenicity studies is advisable, since allergic reactions from fungi might occur. If the MMP is a novel food and a substantial risk is identified during the risk assessment, an indication of allergenicity will be required among the conditions of use in the novel food authorization.

4.5 The Use of Meat-Like Denominations for non-animal based products

MMPs are often used in the production of meat analogues and alternative proteins. One of the most debated labelling issues concerning meat analogues is the use of 'meaty' wordings and denominations when referring to products not containing animal proteins. In the EU, this issue is particularly relevant due to the opposition of certain stakeholders group and governments, such as Italy and France.[17] Both countries have adopted laws to restrict the use of meat-like denominations.[18] The legitimacy of these pieces of legislation can be questioned since the labelling of food products is harmonized at the EU level and the introduction of national prohibitions could jeopardise the free movement of products inside the internal market.

Over the years, attempts have been made to address the matter at the EU level. In 2019 the European Parliament voted down a proposal to prohibiting the use of

15 Selby (2017).

16 Enough Food (2020); Southey (2020).

17 Planchestainer (2024).

18 LEGGE 1° dicembre 2023, n. 172; Décret n° 2022-947 du 29 juin 2022; Décret n° 2024-144 du 26 février 2024.

meat-like denominations for products not containing animal proteins.[19] However, a recent judgement of the CJEU has offered new insights on the topic.

In October 2024, the Court of Justice of the EU (CJEU) clarified that in the EU, in the absence of EU and national laws determining their legal names, the use of words concerning the shapes of products traditionally containing animal proteins is generally permitted for plant-based alternatives, within the limits of the prohibition of misleading the consumer.[20] This encompasses, for example, "veggie burgers" or "plant-based sausages." However, following a previous judgment concerning milk and dairy alternatives,[21] the CJEU further specified that, since EU law defines several types of meats, these cannot not be used to label alternatives. For example, EU law specifies what "meat" is and what "chicken meat" means.[22] Therefore, the use of names like "plant-based chicken" or "vegan beef" is not allowed. According to a number of judgments in Member States, however, names which clearly indicate the "alternative" character of the product, such as "cheese alternative," are allowed, as these names are not likely to mislead consumers.[23]

The CJEU has thus clarified that Member States are not permitted to impose prohibitions on the use of shape-like denominations for non-animal products, as these relate to customary and descriptive names.

Member States are only allowed to enact laws establishing new legal names under Article 38 of the FIC. In practice, this means that countries such as Italy or France can, in principle, introduce legally binding definitions for meat products, provided these definitions specify the actual composition of the products and delineate the boundaries of legal names. For instance, national laws could define traditional preparations like "schnitzel" or "brasato," specifying the ingredients and characteristics required for these products. This could prevent their generic use for non-animal-based alternatives, such as "plant-based brasato" or "vegan Schnitzel."

Depending on how the judgment is received by Member States, it is conceivable that governments opposed to alternative proteins like MMPs, may attempt to establish legal names for specific meat preparations. In cases where a legal name is defined, its use for products that do not meet the specified criteria would be prohibited, even if accompanied by clarifications regarding the substitution of components or ingredients. Over time, this could hinder companies in the alternative protein sector from using terms that describe not only the nature of their products but also their shape, potentially reducing consumer acceptance of these innovative products.

The ability to establish legal names is not without legal limitations. According to Article 4(2) of the FIC, such laws must consider "a widespread need on the part of the majority of consumers for certain information to which they attach significant value, or any generally accepted benefits to the consumer." This means that national

[19] European Parliament (2021).

[20] *C-438/23 (2024).*

[21] *C-422/16 (2017).*

[22] Regulation (EC) No 853/2004, Annex I, Definitions.

[23] Purnhagen (2025).

laws establishing legal names must demonstrate that they contribute to helping consumers make informed choices.

This may explain why the CJEU has emphasized that the right of Member States to establish legal names is limited to situations when certain practices would mislead the average consumer.[24] For instance, one could argue that an EU law requiring foods to be labelled solely according to traditional characteristics, in cases where this prevents consumers from identifying products in line with their expectations, may not comply with Article 7 of the FIR and, by extension, Article 4 of the FIR.

For example, if substitute products—clearly identifiable to the average consumer by shape, taste, and design (such as well-designed mycelium bacon)—are required to bear names that obscure their nature (e.g., "mycelium-derived sliced product"), such a law could eventually be deemed misleading to consumers.

4.6 Clean Labels and Voluntary Claims on Mushrooms and Mycelium Products

Clean labels refer to labels that list only a few, easily recognizable ingredients prepared using traditional methods, often accompanied by claims emphasizing the perceived naturalness and health benefits of the product. They align with a growing consumer preference for avoiding highly processed food products that contain numerous ingredients, colorants, aromas, and flavourings, particularly those identified by unfamiliar numbers or lengthy chemical names. In the broadest sense, clean labels reflect shifting consumer perceptions, particularly in Western countries, and the increasing demand for healthy, natural, and sustainable products.[25] In practice, clean labels provide consumers with the impression that the food they are consuming is free from unnatural or unhealthy components, while conveying that the product possesses desirable qualities.[26] Examples of claims often used in clean labels include: "plant-based," "no GMOs," "made with all natural ingredients," "good for the environment," but also "vegan," "free from palm oil," and "organic."

All this information is voluntarily provided by food business operators. Such voluntary information has to adhere to the requirements of Article 36 FIR. According to this provision they are required, at the very least, to not be misleading and based on the relevant scientific data. In some cases, they need to comply with stricter rules established at the EU or national level.

In the remainder of the chapter, we examine various types of claims and their conditions of use, with a particular focus on the "plant-based" claim, nutrition and health claims, vegan and vegetarian designations, the concept of organic labelling, and the use of environmental and green claims.

---

[24] *C-438/23 (2024).*

[25] Asioli et al. (2017).

[26] Asioli et al. (2017).

4.7 Plant-Based or Fungi Based?

Depending on the food business operator's marketing strategy, it may be attractive to label mycelium-based products as "plant-based." In the EU, there is no legal definition for the claim 'plant-based.' In the absence of mandatory provisions, the rules on voluntary labelling of Art. 36 FIC apply;[27] hence the term "plant-based" needs to fulfil the requirements of Art. 36 FIC. Such a label must not mislead the average consumer,[28] which means that it must be accurate according "to the characteristics of the food." [29] Mycelium-based products are not 'plants,' since they are part of the kingdom Fungi and not Plantae. The use of the term 'plant-based' for fungi products hence does not mirror the characteristics of the product and should hence be considered misleading. The same argument can be derived from the requirements of Art. 36 (2) (c) FIC, which requires that voluntary information is based "on the relevant scientific data." Science has long classified fungi as separate from plants. The use of the claim "plant-based" should thus be avoided when the final product is primarily constituted by the biomass of the mycelium itself or by proteins extracted from it. Wordings like 'fungi-based' or 'mushroom-based' seem more suitable to avoid any possible confusion on the part of the consumer.

4.8 Nutrition and Health Claims

Specific legislation exists in EU food law which allows food business operators to offer consumers voluntary information, such as nutrition and health claims.

Per Regulation (EC) No 1924/2006 on nutrition and health claims, "claim" means any message or representation that is not mandatory under EU law, including graphic, pictorial or symbolic representations which imply that a product has some particular characteristics.[30] This entails not only the label of the product but also advertisement materials, webpages or TV spots.

Nutrition and health claims are two separate categories of claims. Nutrition claims focus on the nature of the products, while health claims are based on the effects the products have. In other words, nutrition claims are related to the product composition from a nutritional point of view, e.g. "high in protein".[31] Health claims instead imply a connection between the product composition and health. For example, the claim "essential fatty acids are needed for normal growth and development

[27] Purnhagen and Molitorisová (2022).

[28] Regulation (EU) No 1169/2011, Article 36 (2) (a) FIC.

[29] Regulation (EU) No 1169/2011, Art. 7 (1) (a) FIC.

[30] Regulation (EC) No 1924/2006, Article 2(2)(1).

[31] Regulation (EC) No 1924/2006, Article 2(2)(4).

of children".[32] Health claims must not suggest that a human disease is treated cured or prevented by eating a food.

Authorized nutrition claims are listed in the Annex of Regulation (EC) No 1924/2006. Anyone can use them, provided that their products meet the specified conditions. A product can bear the claim "high in protein" if 20% of its energy value is provided by protein, while the claim "high in fibre" requires at least 6 g of fibre per 100 g of product, or at least 3 g of fibre per 100 kcal. Health claims instead shall be authorized first,[33] and be included in the Union Register, where the conditions of use are specified.[34] In addition, health claims shall always be accompanied by other mandatory particulars.[35] For example, "Betaine contributes to normal homocysteine metabolism" can only be used for food which contains at least 500 mg of betaine per quantified portion and shall bear the indications: "the beneficial effect is obtained with a daily intake of 1.5 g of betaine" and "a daily intake in excess of 4 g may significantly increase blood cholesterol levels." Data supporting nutrition and health claims should always be available. B2B companies should consider that their customers might require evidence to substantiate nutrition and health claims in the final products.

Nutrition and health claims could be theoretically made only on those foods that meet a nutrient profile awaiting to be prepared by the European Commission under Regulation (EC) No 1924/2006.[36] Nutrient profiles were supposed to avoid the use of nutrition and health claims on products matching their conditions of use but otherwise categorised as unhealthy. However, such nutrient profiles have never been adopted.

---

[32] Regulation (EC) No 1924/2006, Article 2(2)(5).

[33] Regulation (EC) No 1924/2006, Articles 15, 16, 17.

[34] Regulation (EC) No 1924/2006, Article 20(2). The EU Register for Nutrition and Health claims can be found at the following link: https://ec.europa.eu/food/safety/labelling_nutrition/claims/register/public/?event=register.home. Health claims authorised on the basis of proprietary data are recorded in a separate Annex to the Register: https://ec.europa.eu/food/safety/labelling_nutrition/claims/register/public/?event=getListOfPropClaims.

[35] Regulation (EC) No 1924/2006, Article 10(2). In particular, a statement on the importance of a varied and balanced diet and a healthy lifestyle, the quantity of the food and the pattern of consumption required to obtain the beneficial effect; a statement addressed to persons who should avoid the food, if applicable; an appropriate warning for products that are likely to present a health risk.

[36] Regulation (EC) No 1924/2006, Article 4(1).

4.9 Vegan/Vegetarian

There is no legal definition of "vegan" and "vegetarian" in EU law,[37] despite an obligation for the Commission to adopt proper rules in this respect[38] and some stakeholders lobbying for them.[39] Some Member States define "vegan" or "vegetarian" in their national legislation.[40] In the absence of legal definition, private bodies provide independent certification for vegetarian and vegan products, via associations like the European Vegetarian Union.[41]

Like all voluntary claims, vegan and vegetarian claims— even when certified by a private entity—are subject to general labelling rules, in particular Article 36 FIC,[42] as already discussed. MMP may be considered vegetarian whenever they are not accompanied by or contain ingredients derived from meat, and vegan if they are free from ingredients of animal origin; "free" meaning not exceeding a certain predefined threshold. Production processes, the use of additives, potential contaminations must also be taken into account, as it is specified for example in private standards developed by the European Vegetarian Union.[43]

A potential path to label products as vegan/vegetarian is to work with recognized partners and seek appropriate certification under trustworthy private schemes. Evidence must be appropriately recorded to prove that the claims are not misleading.

4.10 Organic Certification and Labelling

Organic production is regulated by Regulation (EU) No 848/2018 on organic production and labelling of organic products. It is defined as an agricultural method that aims to produce food using natural substances and processes, minimizing the environmental impact and encouraging the responsible use of soil, water and resources, the protection of biodiversity and the preservation of ecological balance.[44]

If production requirements set by Regulation (EU) No 848/2018 are respected throughout the entire supply chain,[45] and companies are certified by accredited

---

[37] Sochirca (2018).

[38] Regulation (EU) No 1169/2011, Article 36(3)(b).

[39] Michail (2021); European Vegetarian Union (2021).

[40] Gerke and Janssen (2017).

[41] European Vegetarian Union (n.d.).

[42] Purnhagen and Molitorisová (2022).

[43] European Vegetarian Union (2019).

[44] Regulation (EU) 2018/848, Articles 4, 5.

[45] Regulation (EU) 2018/848, Article 2(2) and Annex II Part I. The Annex specifies rules applicable to Plant Production, the category under which MMP would fall. In particular see Sect. 2.1, in which requirements applicable to the substrates for mushrooms' production are specified.

control bodies and authorities, companies can label their products as organic and use the organic logo on the packaging of products.[46]

Organic production must take place on the soil, and therefore, some innovative production systems, such as vertical farming, cannot be certified as such.[47] However, the regulation also applies to certain products closely linked to agriculture, including edible mushrooms. This encompasses the entire fungal body as well as the mycelium.[48] Substrates for mushrooms and mycelia growth are considered a "reproductive material" production.[49] The use of substrates for mushroom production is permitted as long as these substrates are organic and respect certain conditions.[50] In the case of companies which employ such organic substrates and then sell the final product as organic, it is important to work with trustworthy suppliers, keep records of certification and certify the final product. The same is valid for those companies which acquire organic-produced MMP from third parties to include them in their own final products.

When applying for organic certification, several key aspects must be taken into account. First, GMOs are strictly prohibited in organic production. Additionally, only additives, flavourings, and processing aids specifically approved for use in organic agriculture are permitted.[51] Second, a conversion period to modify production processes from conventional to organic is granted to food business operators willing to engage with organic production.[52] Finally, Regulation (EU) No 848/2018 terminated all recognitions of equivalences previously granted to 'organic' labelling schemes in third countries and requires compliance with EU rules for organic

[46] Regulation (EU) 2018/848, Article 33. The logo can only be used on products if they contain at least 95% of organic ingredients and additionally respect further strict conditions for the remaining 5%. The same ingredient cannot be present in organic and non-organic form. Single ingredients can still be labelled as organic, only in the list of ingredients, if the organic percentage is in total less than 95% (Article 30). Following the same approach held for information and claims, a product is considered to be labelled as organic if specific terms or wording e.g. bio are used, either on the labels or in every other accompanying materials.

[47] Kuljanic (2022), Politico (2021).

[48] European Commission (2024).

[49] European Commission (2024).

[50] Regulation (EU) 2018/848, Annex 2.1. Rules on mushroom production For the production of mushrooms, substrates may be used if they are composed only of the following components: (a) farmyard manure and animal excrement: (i) either from organic production units or from in-conversion units in their second year of conversion; or (ii) referred to in point 1.9.3, only when the product referred to in point (i) is not available, provided that that farmyard manure and animal excrement do not exceed 25 % of the weight of total components of the substrate, excluding the covering material and any added water, before composting; (b) products of agricultural origin, other than those referred to in point (a), from organic production units; (c) peat, not treated with chemical products; (d) wood, not treated with chemical products after felling; (e) mineral products referred to in point 1.9.3, water and soil.

[51] Regulation (EU) No 2018/848, Article 5 (iii) and Article 16. GMOs for veterinary use are permitted.

[52] Regulation (EU) No 2018/848, Article 10.

products imported from third countries.[53] This is particularly relevant for organic producers of MMP planning to export their products to the EU and sell them as organic.

4.11 Environmental Claims: The Unfair Trade Practice Directive and the Proposal for a Green Claims Directive

Green claims, or environmental claims, are assertions made by producers, traders, manufacturers and retailers about the environmentally beneficial aspects of their products.[54] Green claims thus fall under the category of voluntary information. Examples of such claims include reduction of greenhouse gas emissions, protection of endangered species and general information related to a company or organization's positive environmental performance.[55] Currently, the majority of green claims are based on standards developed by private entities, such as industry consortiums, consumer associations or private citizens initiatives, that aim at raising the market value of products and fulfilling consumers' desire for making sustainable choices.

The general rules on voluntary information, in particular Article 36 FIR, apply to green and environmental claims.[56] Most of these claims are based on compensation models; hence on complex data assessing the environmental impact and its compensation. Thus, the assessment of whether they are based "on the relevant scientific data" according to Art. 36 (2) (c) FIC is of major importance to determine whether they are prohibited by the law. In recent years, however, the EU legislator has begun a broader process of harmonization concerning environmental claims by adopting specialised rules.

[53] Regulation (EU) No 2018/848, Article 45 to 49. Under the previous organic regulatory framework, recognition to other organic certification systems in third countries was easily granted. As a consequence, organic producers in a third country could import in the EU and label their product as organic if they were certified as such in the country of origin. All these recognitions expired in 2022. This means that in the absence of a new recognition agreement, producers in third countries now must comply with Regulation (EU) No 848/2018 and be certified by control bodies recognized by the EU.

[54] OECD Committee on Consumer Policy (2010).

[55] For an overview of the existing private standards, see 'Standards Map' <https://standardsmap.org/> To be included in the Standards Map, standards need to address at least one of the pillars of sustainable development (social, environmental or economic), have to present a published set of criteria and an implementation system.

[56] Purnhagen (2024).

4.11.1 Amendments to Directive 2005/29/EC

With the adoption of Regulation (EU) 2024/825 on empowering consumers for the green transition through better protection against unfair practices and better information,[57] a definition of environmental claims was included in Directive 2005/29/EC concerning unfair business-to-consumer commercial practices in the internal market which generically regulates commercial practices in the EU. Environmental claims are defined as:

> any message or representation which is not mandatory under Union or national law, in any form, including text, pictorial, graphic or symbolic representation, such as labels, brand names, company names or product names, in the context of a commercial communication, and which states or implies that a product, product category, brand or trader has a positive or zero impact on the environment or is less damaging to the environment than other products, product categories, brands or traders, or has improved its impact over time;[58]

The amendments to Directive 2005/29/EC also specify that making an environmental claim related to future environmental performance without clear, objective, publicly available and verifiable commitments constitutes misleading actions.[59] The amendments will be applicable from 27 September 2026.

The amendments apply horizontally to all products, not only food products, for which the general obligations of the FIC remain the primary framework, as of today. Further efforts to expand the amendments to Directive 2005/29/EC and harmonize the topic of environmental claims in the EU have been taken with a legislative proposal for the specific regulation of green claims.

4.11.2 A proposal for a Green Claims Directive

In addition to the general amendments of Directive 2005/29/EC, on 22 March 2023, the European Commission published a proposal for a Directive concerning the use of environmental claims in the EU, the "Green Claims Directive".[60] The Directive aims at harmonizing and defining rules for the use of so-called "green claims" in the EU. So far, regulation of sustainability labels has largely relied on existing EU competition and internal market law.[61] The proposed framework aims to regulate the use of all claims that concern the environmental performance of products, with the objective of preventing companies from using unclear or unsubstantiated environmental claims and enabling consumers to make informed choices.[62]

[57] Directive (EU) 2024/825.

[58] Directive 2005/29/EC, Article 2 (o).

[59] Directive 2005/29/EC, Article 6(2)(d)

[60] European Commission (2023).

[61] Schebesta (2019).

[62] European Commission (2023). Section 1.1. Reasons for and objectives of the proposal.

The Green Claims Directive's main goal is thus to prevent companies from making environmentally false statements, a phenomenon known as "greenwashing".[63] Green claims should be based on measurable data, similar to nutrition and health claims. Companies would be required to demonstrate that the environmental aspects subject to the claim are significant from a life cycle-perspective, and not equivalent to requirements already imposed to all similar products.[64] The EU Green Claims Directive would also limit the proliferation of private sustainability labels and logos, by allowing their creation only when they provided a significant added value.[65] Since the power of green claims often lies in the impact of their logo, the proposal aims to address the negative response that consumers have when subject to an overload of stimuli.[66]

The proposal for a Green Claim Directive has followed the ordinary legislative procedure in the EU. On 17 June 2024, the European Council adopted its approach to the proposed Directive. Negotiations with the European Parliament began at the end of 2024.

Court Cases

C-195/14 - Teekanne (2015) Court of Justice of the European Union ECLI:EU:C:2015:361

C-210/96 - Gut Springenheide and Tusky v Oberkreisdirektor des Kreises Steinfurt (1998) Court of Justice of the European Union ECLI:EU:C:1998:369

C-422/16 - Verband Sozialer Wettbewerb eV v TofuTown.com GmbH (2017) Court of Justice of the European Union ECLI:EU:C:2017:458

C-438/23 - Protéines France and Others (2024) Court of Justice of the European Union *ECLI:EU:C:2024:826*

C-470/93 - Verein gegen Unwesen in Handel und Gewerbe Köln v Mars (1995) Court of Justice of the European Union ECLI:EU:C:1995:224

Instruments and Legislation

European Commission, 'Frequently asked questions on organic rules' (2024) https://agriculture.ec.europa.eu/system/files/2023-11/organic-rules-faqs_en.pdf.

[63] Evans (2024).

[64] European Commission (2023). Article 3.

[65] European Commission (2023). Section 1.1. Reasons for and objectives of the proposal and Sect. 3.2.1. Problem definition & preferred policy option.

[66] Moon et al. (2017); see for example Rainforest Alliance (2017).

European Commission, 'Proposal for a Directive of The European Parliament And Of The Council on substantiation and communication of explicit environmental claims (Green Claims Directive)' (2023) COM/2023/166

European Parliament, 'Are veggie burgers, tofu steaks or the use of yogurt pots going to be banned? | News | European Parliament' (2021), available at https://www.europarl.europa.eu/news/en/press-room/20201019BKG89682/eu-farm-policy-reform-as-agreed-by-the-parliament-and-council/7/are-veggie-burgers-tofu-steaks-or-the-use-of-yogurt-pots-going-to-be-banned.

LEGGE 1° dicembre 2023, n. 172. Disposizioni in materia di divieto di produzione e di immissione sul mercato di alimenti e mangimi costituiti, isolati o prodotti a partire da colture cellulari o di tessuti derivanti da animali vertebrati nonché di divieto della denominazione di carne per prodotti trasformati contenenti proteine vegetali.

OECD Committee on Consumer Policy, 'Environmental claims' (2010), available at https://www.oecd.org/sti/consumer/48127506.pdf

Regulation (EC) No 1924/2006 of the European Parliament and of the Council of 20 December 2006 on nutrition and health claims made on foods, OJ L 404, 30.12.2006 p. 9–25

Regulation (EC) No 853/2004 of the European Parliament and of the Council of 29 April 2004 laying down specific hygiene rules for food of animal origin, OJ L 139, 30.4.2004, p. 55–205

Regulation (EU) 2015/2283 of the European Parliament and of the Council of 25 November 2015 on novel foods, amending Regulation (EU) No 1169/2011 of the European Parliament and of the Council and repealing Regulation (EC) No 258/97 of the European Parliament and of the Council and Commission Regulation (EC) No 1852/2001, OJ L 327, 11.12.2015, p. 1–22

Regulation (EU) 2018/848 of the European Parliament and of the Council of 30 May 2018 on organic production and labelling of organic products and repealing Council Regulation (EC) No 834/2007, OJ L 150, 14.6.2018, p. 1–92

Regulation (EU) No 1169/2011 of the European Parliament and of the Council of 25 October 2011 on the provision of food information to consumers, amending Regulations (EC) No 1924/2006 and (EC) No 1925/2006 of the European Parliament and of the Council, and repealing Commission Directive 87/250/EEC, Council Directive 90/496/EEC, Commission Directive 1999/10/EC, Directive 2000/13/EC of the European Parliament and of the Council, Commission Directives 2002/67/EC and 2008/5/EC and Commission Regulation (EC) No 608/2004, OJ L 304, 22.11.2011, p. 18–63

Commission Implementing Regulation (EU) 2017/2470 of 20 December 2017 establishing the Union list of novel foods in accordance with Regulation (EU) 2015/2283 of the European Parliament and of the Council on novel foods, C/2017/8878, OJ L 351, 30.12.2017, p. 72–201

Décret n° 2022-947 du 29 juin 2022 relatif à l'utilisation de certaines dénominations employées pour désigner des denrées comportant des protéines végétales;

Décret n° 2024-144 du 26 février 2024 relatif à l'utilisation de certaines dénominations employées pour désigner des denrées comportant des protéines végétales.

Directive (EU) 2024/825 of the European Parliament and of the Council of 28 February 2024 amending Directives 2005/29/EC and 2011/83/EU as regards empowering consumers for the green transition through better protection against unfair practices and through better information, PE/64/2023/REV/1, OJ L, 2024/825, 6.3.2024.
Directive 2000/36/EC of the European Parliament and of the Council of 23 June 2000 relating to cocoa and chocolate products intended for human consumption, OJ L 197, 3.8.2000, p. 19–25
Directive 2005/29/EC of the European Parliament and of the Council of 11 May 2005 concerning unfair business-to-consumer commercial practices in the internal market and amending Council Directive 84/450/EEC, Directives 97/7/EC, 98/27/EC and 2002/65/EC of the European Parliament and of the Council and Regulation (EC) No 2006/2004 of the European Parliament and of the Council ('Unfair Commercial Practices Directive'), OJ L 149, 11.6.2005, p. 22–39.

References

Bibliography

Asioli D et al (2017) Making sense of the "Clean Label" trends: a review of consumer food choice behavior and discussion of industry implications. Food Res Int 99:58
Brennan D (2021) What to know about mycoprotein. Nourish. Available at https://www.webmd.com/diet/what-to-know-about-mycoprotein
Charette P, Hooker NH, Stanton JL (2015) Framing and naming: a process to define a novel food category. Food Quality Pref 40:147
Enough Food 'What Is It?' (2020). Available at https://www.enough-food.com/what-is-it
European Vegetarian Union (2019) Definitions of "Vegan" and "Vegetarian" in Accordance with the EU Food Information Regulation. Available at https://www.euroveg.eu/wp-content/uploads/2021/02/072019_EVU_PP_Definition.pdf
European Vegetarian Union (2021) Vegan and Vegetarian Definitions. Available at https://www.euroveg.eu/vegan-and-vegetarian-definitions/
European Vegetarian Union (n.d.) V-Label. Available at https://www.euroveg.eu/v-label/
Evans L (2024) The green claims directive: the future legal framework for advertising with environmental claims. Eur Food Feed Law Rev 19:174
Gerke M, Janssen M (2017) Vegan foods: labelling practice. ErnährungsUmschau 64:51
Hallman WK, Hallman WK (2020) An empirical assessment of common or usual names to label cell-based seafood products. J Food Sci 85:2267
Kuljanic, N (2022) European Parliament think-thank: what if we grew plants vertically? Available at https://www.europarl.europa.eu/thinktank/en/document/EPRS_ATA(2022)737130
Michail N (2021) EU to set legal definition of vegetarian and Vegan Food. Food Navigator. Available at *https://www.foodnavigator.com/Article/2017/11/03/EU-to-set-legal-definition*-of-vegetarian-and-vegan-food
Moon S-J, Costello JP, Koo D-M (2017) The impact of consumer confusion from eco-labels on negative WOM, distrust, and dissatisfaction. Int J Advert 36:246
Planchestainer F (2024) Meat me in Italy: the Italian Ban on sounding names and cell-cultured meat. Eur Food Feed Law Rev 19:66

Politico (2021) Vertical Farming's Sky-High Ambitions Cut Short by Eu Organic Rules. https://www.politico.eu/article/vertical-farming-eu-organic-rules-startups/

Purnhagen K (2024) Klimaneutral und Irreführung – Was ein Blick ins EU-Recht bringen. Europäische Zeitschrift für Wirtschaftsrech 35:738

Purnhagen K (2025) Misery acquaints a man with strange bed-fellows? Consumer empowerment in EU agricultural law and EU food information law. In: Cardwell M, Smith F, McMahon J (eds) A research Agenda for agricultural law. Edward Elgar, Cheltenham

Purnhagen KP, Molitorisová A (2022) Public and private enforcement in European Union food law. Eur J Risk Regul 13:464

Rainforest Alliance (2017) The Rainforest Alliance and UTZ to Merge, Forming New. Stronger Organization. https://www.rainforest-alliance.org/articles/rainforest-alliance-utz-merger

Schebesta H (2019) Control in the label: self-declared, certified, accredited? In: Rott P (ed) Certification – trust, accountability, liability. Springer

Selby G (2017) Quorn Explains Mycoprotein "Mold" Origin as US Labeling Case Is Settled. https://www.foodingredientsfirst.com/news/quorn-explains-mycoprotein-mold-origin-as-us-labeling-case-is-settled.html

Sochirca N (2018) The European legal framework on vegan and vegetarian claims. Eur Food Feed Law Rev 13:514

Southey F (2020) Allergen-Free Fungi Protein a "Breakthrough" in Food Ingredients: "We Have a Unique Proposition from a Sustainability and Economic Perspective". Food Navigator. https://www.foodnavigator.com/Article/2020/11/04/The-Protein-Brewery-develops-breakthrough-allergen-free-fungi-protein-for-food

Chapter 5
Conclusions

This book has illuminated the evolving landscape governing the development, placing on the market, and post-market surveillance of products derived from mushrooms and mycelium. From their classification as novel foods to the specific provisions for food enzymes, additives, and labelling, mushroom and mycelium-based products occupy a pivotal position at the intersection of scientific innovation, consumer expectations, and regulatory oversight.

A recurring theme throughout this book is the balance between fostering innovation, ensuring consumer safety and compliance with established regulatory standards. The EU's rigorous framework, while sometimes presenting challenges for food business operators, reflects the Union's commitment to maintaining high standards for food safety, environmental sustainability, and consumer trust.

As this field continues to grow, it is essential for stakeholders—including regulators, producers, and consumers—to engage collaboratively. Producers must invest in robust research and data to meet regulatory requirements, while policymakers should ensure that regulations remain adaptive to emerging scientific evidence and market trends. At the same time, consumer education will play a vital role in shaping the market acceptance of mushroom and mycelium-based products, particularly as clean labels, plant-based claims, and sustainability considerations become more central to purchasing decisions.

Looking ahead, the regulatory environment must also consider the challenges posed by globalization, technological advancements, and the need for harmonized standards that enable seamless market access while preserving cultural and regional diversity in food production. Addressing these complexities will be crucial to unlocking the full potential of mushroom and mycelium as sustainable, nutritious, and innovative contributors to the global food system.

A. Molitorisová et al., *Regulation of Mushroom- and Mycelium-Based Products in the EU*, SpringerBriefs in Law, https://doi.org/10.1007/978-3-031-93780-4_5

The manufacturer's authorised representative in the EU is Springer Nature Customer Service Centre GmbH, Europaplatz 3, 69115 Heidelberg, Germany. If you have any concerns regarding our products, please contact ProductSafety@springernature.com

Printed and bound by CPI Group (UK) Ltd, Croydon, CR0 4YY
30/09/2026
02225575-0011